Alchemy

Unlocking Secrets of an Ancient Mystical Science

© Copyright 2021

The content contained within this book may not be reproduced, duplicated or transmitted without direct written permission from the author or the publisher.

Under no circumstances will any blame or legal responsibility be held against the publisher, or author, for any damages, reparation, or monetary loss due to the information contained within this book, either directly or indirectly.

Legal Notice:

This book is copyright protected. It is only for personal use. You cannot amend, distribute, sell, use, quote or paraphrase any part, or the content within this book, without the consent of the author or publisher.

Disclaimer Notice:

Please note the information contained within this document is for educational and entertainment purposes only. All effort has been executed to present accurate, up to date, reliable, complete information. No warranties of any kind are declared or implied. Readers acknowledge that the author is not engaging in the rendering of legal, financial, medical or professional advice. The content within this book has been derived from various sources. Please consult a licensed professional before attempting any techniques outlined in this book.

By reading this document, the reader agrees that under no circumstances is the author responsible for any losses, direct or indirect, that are incurred as a result of the use of information contained within this document, including, but not limited to, errors, omissions, or inaccuracies.

Your Free Gift (only available for a limited time)

Thanks for getting this book! If you want to learn more about various spirituality topics, then join Mari Silva's community and get a free guided meditation MP3 for awakening your third eye. This guided meditation mp3 is designed to open and strengthen ones third eye so you can experience a higher state of consciousness. Simply visit the link below the image to get started.

https://spiritualityspot.com/meditation

Contents

INTRODUCTION 1
CHAPTER 1: THE PRINCIPLES OF ALCHEMY 3
    HISTORY OF ALCHEMY 5
    ALCHEMY DURING THE MEDIEVAL PERIOD 6
    ALCHEMY DURING THE RENAISSANCE PERIOD 6
    MODERN ALCHEMY AND ITS CONTRIBUTION TO YOUR LIFE 7
    HEALING AND MEDICINE 8
    PSYCHOLOGY 10
    MODERN OCCULT TRADITIONS 10
    ADDITIONAL NOTEWORTHY INTERPRETATIONS AND CONTRIBUTIONS 11
    THE THREE PRINCIPLES OF ALCHEMY 13
    OTHER VITAL COMPONENTS OF ALCHEMY 17
CHAPTER 2: THE FOUR ELEMENTS AND THE QUINTESSENCE 18
    HOW THE FOUR ELEMENTS/ROOTS EXISTED 19
    THE FOUR ELEMENTS AND THE SYMBOLS THAT REPRESENT THEM 21
    KEY QUALITIES OF THE ELEMENTS 25
    HOW TO BALANCE THE ELEMENTS 26
    WHAT IS QUINTESSENCE? 29
    BRING THE ALCHEMICAL ELEMENTS INTO YOUR LIFE 30
CHAPTER 3: THE PLANETARY METALS 35
    HOW TO TRANSFORM YOUR LIFE WITH THE PLANETARY METAL LEAD 38

How to Transform Your Life with the Planetary Metal Tin 40
How to Transform Your Life with the Planetary Metal Iron 41
How to Transform Your Life with the Planetary Metal Gold 43
How to Transform Your Life with the Planetary Metal Copper 45
How to Transform Your Life with the Planetary Metal Quicksilver (Mercury) 47
How to Transform Your Life with the Planetary Metal Silver 49

CHAPTER 4: THE MUNDANE ELEMENTS 51
CHAPTER 5: THE TWELVE OPERATIONS 1 – 6 62
CHAPTER 6: THE TWELVE OPERATIONS 7 - 12 72
CHAPTER 7: HERBAL ALCHEMY 84

Planetary Metals in Plants and Herbs 85
Plants/Herbs with Planetary Metal Gold 89
Plants/Herbs with Planetary Metal Silver 91
Plants/Herbs with Planetary Metal Iron 93
Plants/Herbs with Planetary Metal Quicksilver (Mercury) 95
Plants/Herbs with Planetary Metal Tin (Jupiter) 97
Plants/Herbs with Planetary Metal Copper (Venus) 99
Plants/Herbs with Planetary Metal Lead (Saturn) 101

CHAPTER 8: THE SPAGYRIC PROCESS 103

Spagyric as Part of Herbal Alchemy 103
Make Spagyric Tinctures and Substances 104
Find Plants and Other Supplies for the Spagyric Alchemical Operation 106

CHAPTER 9: AN ALCHEMICAL COOKBOOK 113

Herbal Infusions 114
Prepare Herbal Infusions 114
Additional Notes and Tips 116
Herbal Tinctures 117
Herb-Infused Oil 119
Other Herbal Alchemy Ideas That You Can Prepare at Home 121
Basic Alchemy Oil 121
Frankincense or Myrrh Tincture 122

CHAPTER 10: BECOMING A MODERN ALCHEMIST 124

CONCLUSION 128
HERE'S ANOTHER BOOK BY MARI SILVA THAT YOU MIGHT
LIKE 129
YOUR FREE GIFT (ONLY AVAILABLE FOR A LIMITED TIME) 130
REFERENCES 131

Introduction

Do you want to unlock the secrets of alchemy, an ancient mystical science? Then prepare every aspect of yourself, including your physical, mental, and emotional faculties, for the mind-blowing discoveries that you will encounter in this book.

The study of alchemy is full of secrets, but it has left a lasting, deep impression on almost all cultures and the hearts and minds of generations.

Alchemy also continues to gain increased recognition as a vital part of chemistry's heritage. It plays a major role in people's continuous desire and attempt to control and explore the natural world and use it wisely. The magnetic appeal of alchemy is most often attributed to how it reaches into the hearts of many and promises to fulfill even their most impossible dreams and desires; anyone can be an alchemist.

If you want to find easy steps to start taking advantage of the power of alchemy, then *Alchemy: Unlocking Secrets of an Ancient Mystical Science* will lead the way. You can find simple interpretations for even the most complex alchemical symbols and learn how to practice it at home.

What else? This book has the most up-to-date information about alchemy. This means that the outdated systems that no longer work are not included in this book. It only has the juiciest and most valuable information that is truly applicable to practicing modern alchemy.

Use this book to learn everything about alchemy and how to practice it to achieve your heart's desires.

Let's start!

Chapter 1: The Principles of Alchemy

Alchemy refers to a famous ancient practice that is cloaked in secrecy. It is a popular yet multilayered and complex subject that serves as one of the earliest forms of chemical technology, exploring the nature of substances.

It is a known art and science that aims to let base metal evolve into gold through refining, purifying, and separating—among other steps and stages. The evolution is necessary for forming the philosopher's stone, which refers to a magical ingredient that serves as a catalyst to convert lead into gold.

The people who first practiced alchemy primarily used it to transform lead into gold and to hasten the process. It resulted in a quest that captured the attention and imaginations of many people over several centuries. However, the main purpose of practicing alchemy goes a lot further than just creating gold.

Alchemy turned into a philosophy about the cosmos, similar to astrology, which is still practiced today. It even founded incredible language symbols and representations that are useful to explore everything about the world.

Alchemy has a strong and solid philosophical basis, which is why it is now integrated into different fields. Alchemists even form alchemical ideas by integrating spiritual matters and religious metaphors into their work.

Another important fact about alchemy is that it comes from a complicated spiritual worldview, where everything surrounding humans takes the form of a universal spirit. Alchemy practitioners even believed that metals are alive and capable of growing within the Earth.

Whenever a common or base metal like lead was discovered in the past, alchemists believed that it was just a physically and spiritually immature variation of a higher metal, like gold. They also believed that metals are not just unique substances forming part of the periodic table, but rather that all of them are similar items in various developmental or refinement stages that are moving their way toward achieving spiritual perfection.

Another perspective about alchemy that many people believe in is that it is a mysterious practice performed by those who are part of a secret society who use codes to exchange messages. Some even view alchemists as holding paranormal powers, giving them the ability to perform magical acts and enjoy an extremely long life. It is the reason why a lot of alchemists are viewed as magicians.

Despite the many points of view regarding alchemy, one thing is for sure—it has made a huge impact in the lives of many. If you are interested in learning more about it, you have to be willing to open your mind and accept more than just stereotypical perspectives about this practice. By opening your mind, you can completely appreciate the different ways it has encouraged and inspired many people for so many centuries.

History of Alchemy

According to what most people believe, the word alchemy originated from "qem" or "chem," an Egyptian term meaning "black." It signifies the black alluvial soils, a specific form of sediment composed of various materials that served as the banks of the Nile. The concept of alchemy began with a strong interest in the material substances present in this world. Those who practiced it tried to find the ultimate wisdom and knowledge.

The evolution of this practice is traceable from the time it started in Egypt and throughout the Middle Ages and Renaissance period. Each one of the periods mentioned brings out a unique flavor of what alchemy denotes and focuses on.

The first time people discovered and heard about alchemy, though, was during the Hellenistic Egypt period. Here, Alexandria was considered as the city where alchemy thrived, making the city its heart. It was during 300 A.D. when Zosimos, an Egyptian, was recognized as one of the most renowned alchemists. It was the reason why his work also got discovered and recognized, eventually leading to Al-Tughra'i, a royal calligrapher, translating his work during 1100 AD into Arabic.

The researcher H. S. El Khadem translated the text in the paper *A Translation of a Zosimos' Text in an Arabic Alchemy Book* (1996), and explored relevant information about the early practitioners of alchemy. The text and work of Zosimos was then further translated into other languages, including English.

With the many languages into which it was translated, new audiences discovered the existence of this practice. They also discovered that alchemical literature has a mystical and mysterious nature aside from it being a form of science. The text offered clear and detailed instructions regarding the philosophies and processes involved in alchemy.

Alchemy During the Medieval Period

As part of alchemy's history, it is necessary to tackle how this practice was continued during the medieval period. It was when the entirety of Europe began having a strong interest in everything related to alchemy. This is because Islam introduced it to Europe after using the practice and saw that is was viable. The world of Islam encouraged alchemy to reach Europe through Spain during the 5th century.

Paracelsus, a famous philosopher during this period, also acted as an alchemist. He practiced alchemy with a focus on health, wellness, and medical practices. He wanted to improve medical practices by retaining the balance between physical and spiritual aspects. As an alchemist, he made it a point to use the knowledge and wisdom he acquired, particularly about transmutation and metals, to treat physical and mental illnesses.

During the Middle Ages alchemy became even more popular. With the many experiments conducted at that time, practitioners started focusing heavily on witchcraft and magic. Since the church prohibited practicing magic and witchcraft, alchemists and scholars secretly conducted their research, and cryptic codes were developed to document the work.

Alchemy During the Renaissance Period

The Renaissance seemed to be the period when alchemy was finally liberated from being a secret practice. This liberation started in the 14th century, mainly because the world had become more interested in rational thought, the changes in perspectives about humans, and the original philosophers who gave birth to and influenced the practice of alchemy.

It was like the practice had been reborn. You could even find alchemists who were part of the pharmaceutical and medical industries. During the Renaissance years, a lot of occultists who were interested in alchemy focused their use of the practice on the state of the soul.

The experiments and research continued with the growing popularity of alchemy. During the 16th to 18th centuries, there were around 4,000 alchemy-related books that were printed and published. Most of these books explored alchemy based on various perspectives.

Its popularity started waning during the 19th century. It was revived once again during the 20th century. Presently, you can see people referring to alchemy as a concept that uses enigmatic and obscure symbols. Some also view it as a concept where the focus is on inner change and spiritual transformation.

Modern Alchemy and Its Contribution to Your Life

At present, you can find many people continuing to practice alchemy and perform alchemical experiments. It is the reason why you can unfold and discover hints of it in various aspects of modern life. You can even see it playing a part in modern science and medicine. It was also discovered that this practice has made significant contributions to various industries, including refining, metalwork, and the manufacture and production of ceramics, ink, glass, gunpowder, liquors, paints, cosmetics, and dyes.

Alchemists were also those who conceptualized the chemical elements that formed part of the first basic periodic tables. Furthermore, they were responsible for the distillation process recognized in Western Europe.

There are many areas where you can find hints of alchemy even during modern times, despite it being an ancient practice.

Healing and Medicine

One area where modern alchemy can have a positive impact on your life is healing. You can still find people who conduct research and experiments related to alchemy, with the goal of producing healing remedies. In this specific field, alchemy focuses more on alternative medicine, because many people in different parts of the world believe in alternative healing methods.

Today, several alternative practices have a history that can easily be traced back to the alchemy practiced in China, India, and the East. Taking that into account, it is safe to say that the practice of alchemy still plays a major role when it comes to gaining a full understanding of the way transformation occurs within the human body. It offers several other healing options apart from just sticking to the holistic approach.

Alchemy looks closely at your mind and body and finds their connection, making it possible for the practice to follow a different approach to healing imbalances and disease. Medicinal, health, and wellness practices that adhere to the philosophies and principles of alchemy are still currently used in Ayurveda.

You can trace such a connection to the alternative medicine of Ayurveda by discovering "rasa," which is a word in Ayurveda texts. This term is the equivalent of qi (life force) in Chinese medicine. If you try to unlock and harness your rasa, this action can give you easy access to a wide range of health benefits and a longer lifespan.

Ayurveda also tackles the power of herbs and how they can positively influence the human body. Apart from that, rasa in Ayurveda also focuses more on preserving life instead of fighting various ailments through medicine. This specific principle closely connects to how alchemists work, especially if you connect it to how they search for eternal youth and immortality and produce various concoctions to achieve it.

You can also see alchemy being closely related to traditional Chinese medicine. As a matter of fact, China founded an alchemical system strongly based on the philosophies of the Tao. These philosophies examine the energy and patterns that form part of the universe. The main purpose of doing so is to harmonize energies, signifying how alchemy in China emphasizes finding balance within the human body.

Apart from that, the Chinese alchemy system heavily emphasizes the need to look for a solution to achieve immortality and longevity. This makes its philosophies and principles quite similar to the alchemical principles practiced by other traditions. A famous ingredient used in concocting the elixir is Cinnabar, which refers to a mercury sulfide capable of balancing your yin and yang.

It is a toxic substance when you ingest it, though, which is why Chinese medicine at present tries to provide balance in the human body without using harmful and poisonous substances. Nowadays, Chinese medicine also tries to cure different forms of disease using safe herbs and acupuncture, all of which have roots in the alchemic system set in place in the past.

It even provides steps to distil substances and produce elixirs designed to promote eternal youth. Moreover, the translated texts provided information about Greek and Aristotelian philosophies that proved the amalgamation of various cultural symbols present in alchemy.

Psychology

You can also find alchemy being integrated into the field of psychology. Psychologist Carl Jung integrated the ideas of alchemy into his theories about the transformation of the human psyche. You will also notice esoteric traditions being completely based on the texts, legends, symbols, and lore used in alchemy.

This practice had a major influence on Jung, whose work strongly focused on actualization and self-transformation. Both of which are achievable by going through the process referred to as individuation. At present, Jungian psychology, which touches on alchemy, consists of practical applications when it comes to treatment and healing as well as in psychedelic therapy settings. Also, Jungian therapists are recognized for their jobs that let their clients improve their ability to manage and handle all their mental energies that are in conflict.

Moreover, many perceived the practice of alchemy as a major component of psychology when viewed on a much deeper level. The people who believed this concept also had faith in alchemy's ability to unlock a clear understanding of one's psyche—whether it was the subconscious or the innermost part of the thought processes.

Note that the psyche is the primary part of your being, so getting a full understanding of it is necessary. Your knowledge about alchemy and practicing it correctly can help you to better understand yourself.

Modern Occult Traditions

Another area where alchemy seems to shine is in implementing modern occult traditions. Hermes, a famous Greek God, and Mercury, a Roman God, played a vital role in forming the esoteric and alchemical texts that greatly influenced the occult traditions used during these present times. Hermes has been known to be capable of traveling freely from one world to another.

He could also travel the world below as well as the Earth and the heavens. This ability can be somehow attributed to the practice of alchemy. This Greek God also carried with him a caduceus. It refers to a staff composed of two snakes entangling around it. This caduceus turned out to be a symbol representing healing, wellness, and medical practices and is commonly used to symbolize medical facilities today.

Many also became interested in using alchemy, as it serves as an incredible source of esoteric and philosophical ideas. If you are interested in this, then you have the freedom to use all the ideas and symbols linked to alchemy to do things like tarot reading or creating a magical or esoteric system.

Additional Noteworthy Interpretations and Contributions

A lot of those who practiced or who were interested in alchemy also viewed it as a popular spiritual and philosophical field of study. It used a combination of chemistry and metalwork, and was also interpreted as a way to investigate nature. It is the reason why you can find alchemy influencing parts of medicine, physics, spiritualism, art, mysticism, and astrology, among many other practices.

Alchemy has a couple of goals that are beneficial for humankind, including the following:

- **Discovering the ultimate elixir of life.** Most practitioners believe that what they will find is a magical elixir designed to bring them not only good health and wealth but also eternal life.

- **To search for or create the substance referred to as "the philosopher's stone."** Once created, it is crucial to heat it then mix it with either iron or copper to form gold, the purest and highest form of matter.

- **Figuring out the human connection to the cosmos.** It is necessary to understand this relationship, as a means of boosting the human spirit.

Considering the goals associated with practicing alchemy, the whole field of study is also spiritual. Some alchemists try to unravel the secret of purifying iron or copper while aiming to turn it into gold. When they do, they use it to improve their ability for human soul purification.

Other alchemists were also serious practitioners who used this field of study and made it part of modern medicine and chemistry. With the many good things that alchemy has contributed to various aspects of life and in major sciences, it is safe to say that its focus nowadays, and even in the past, is self-transformation, instead of the more scientific goals, like metal conversions.

For instance, similar to when the ancient alchemists attempted to transmute or transform lead into gold, you can also use modern mental alchemy to turn or transform your ideas into a reality that you can see and touch. Examples where you can take full advantage of this self-transformation focus include:

- Transmuting pain and hardships into something beautiful and genuinely enjoyable.
- Transforming deep longing and sadness into a passionate romantic relationship.
- Transmuting the sudden surge of inspiration into a flourishing and successful business.
- Transforming body discomforts and unhealthiness into a health-focused lifestyle and your dream body.

There are several things that you can do with alchemy, which is proof of how it can contribute to many areas in your life in a positive way. It gives you the power and ability to transform and transmute almost everything in your life.

This means that whether you want to improve your romantic or family relationships, your career, finances, or acquire something material, you have the power to do it right now. You can even turn extremely difficult situations into something that will let you manifest even those desires that were seemingly impossible in the past.

The Three Principles of Alchemy

Now that you know about alchemy, its major contributions from the past to the present, some of its interpretations, and a brief history, it is time to delve deeper into it by understanding the principles governing it. Alchemy functions based on three philosophical principles. Also called the "tria prima" or "three primes," it is a trinity composed of salt, mercury, and sulfur.

It was Jabir ibn Hayyan who first formulated the understanding of mercury and sulfur. Later on, Paracelsus expanded this understanding, recognizing salt, mercury, and sulfur as the three primary bodies where the four classic elements also existed. According to practitioners, it is possible to divide all matter into the three principles indicated through alchemical processes.

Alchemists also believe that these three principles are the three-fold aspects that govern all phenomena, so you can expect to find them in all organic nature and mineral compounds.

Sulfur

Sulfur is one principle in alchemy, which strongly symbolizes an expanding force. It represents the spirit, transcendence, and essence. It also serves as the soul. Sulfur works by molding you into someone who is balanced and has a strong motivation and desire to take action creatively.

This specific principle has a strong connection to the air and fire elements. You can view it as the principle that gives motion. It is also combustible, which is why it symbolizes the inner desires and ambitions of humans. Moreover, it can be signified as the spiritual nature constructed within.

This principle is all about ambition, desire, and the spirit's fiery nature, which is why alchemists also link it to the sun as well as its inherent qualities. These include being dry and hot. Chemically, you can see it residing within the essential oils that you work with in the laboratory.

Salt

Salt refers to the fixity or non-action principle. It also serves as the principle of incombustibility, which represents the body. It holds the earth element, which symbolizes the vegetable salt that can be extracted from the calcined ashes found in the bodies of plants. As it signifies a thing's body, it is safe to say that salt acts as the receptacle for all the energies that come from sulfur and mercury.

Having a strong connection to both the water and earth, salt also symbolizes the physical world. You may view it as a more constricting force, which is also the opposite of sulfur. Salt represents dry and cold.

It is a fixed principle that means balance as well as the union of male and female. There are even times when salt symbolizes a child. As far as chemical aspects are concerned, the neutral energy is present in the salts found in laboratory materials, whether they are insoluble or soluble.

Mercury

The third substance is mercury, which refers to the principle of volatility and fusibility, an indication of the ability to flow and melt. It has a strong relationship with the water element as well as that of air. It represents the plant's life essence.

Mercury can also be defined as the female and passive energy governing the universe. It acts as a balancing point, making it capable of providing a balance between the spirit or soul and the human body. This means that it is the specific principle which brings the physical and spiritual aspects together, making it possible for them to communicate.

Aside from that, mercury has a strong connection to one's intellect. It corresponds to the mind, knowledge, and wisdom. It is also capable of mixing with sulfur while trying to create a manifestation, regardless of what it is. Chemically, you can see it as a form of energy, which is a part of the alcohol content in the specific material you are working with.

Other Vital Components of Alchemy

Apart from the mentioned substances, sulfur, salt, and mercury, which serve as the three major principles of alchemy, this field is also governed by four essential elements—earth, fire, water, and air. We will discuss these elements in more detail in the next chapter of this book. A lot of alchemists use these elements as their basis, especially in founding alchemical principles and philosophies.

Apart from that, symbols and images are also vital components in alchemy, especially if you consider how alchemists use symbolic imagery and language to push through their transmutations. Among the popular symbols are snakes that signify regeneration, as well as a fire that represents purification. You can see these symbols in numerous alchemical texts, proving the major role they played in this field.

You can also see planetary symbolism being used in alchemy, especially in drawings. Among these symbols are the Moon, Mercury, and Sun. You can even sense the strong relationship between astrology and alchemy. It is the main reason why you can find several elemental and symbolic correspondences that overlap between these two fields.

Chapter 2: The Four Elements and the Quintessence

One thing that gives alchemy a strong and distinctive character is its end goal, which is transmutation. It is the result that you have to achieve whenever you practice alchemy. You can view and comprehend transmutation in several ways, among which are chemical changes as well as physiological changes. One example would be transmuting from sickness to health. Other changes include transforming from old age into being youthful or transmuting from an earthly existence into the supernatural.

The good thing about the transmutations or changes produced by alchemy is that all of them are always positive. The changes never involve degradation, though you may feel some slight discomfort during the intermediate stages of attaining the happy and satisfying ending you are hoping for. All in all, this field focuses more on giving humans everything considered as good, including longevity, immortality, and wealth.

One thing to note is that the ultimate goal of transmutation is only achievable by following all the principles of alchemy. You also need to understand its four major elements: air, earth, fire, and water. What do these elements mean, and how can your full understanding of them help you achieve your desired transformation? Let's get to know more about that in this chapter.

How the Four Elements/Roots Existed

Initially referred to as the four roots according to the Greek philosopher Empedocles, fire, air, water, and earth are the four elements of alchemy. It was Plato who was responsible for classifying them as elements. He was the one who first used the term "elements" when referring to the roots. Others followed suit, and to this day, alchemists use "elements" to refer to the four roots.

The concept of using the four elements was actually responsible for the development of the most vital ideas of future alchemy. These elements work under the principle that a substance's properties are based on its components. This resulted in the belief that making some changes to the properties of a substance will transform it, which is how alchemy actually works.

According to the ancient alchemists, these four elements were also associated with the four significant traits of the physical world. Furthermore, they serve as the building blocks of transformation and creation. Alchemists also believed that these four elements were the significant archetypes within matter. They represent metaphysical qualities. Empedocles also believed that two of these elements—air and fire—referred to the external ones. This meant that you could use them to reach outward and upward. The remaining two—water and earth—worked downward and inward.

Another thing that you have to learn about these four elements is that alchemists do not only consider them as material substances. They also describe them as spiritual essences. They are categorized as archetypes, which means it is crucial to get a full experience of these elements to gain a complete understanding of each one. Empedocles also made a point of connecting every element with a specific god— Hera (earth), Hades (fire), Persephone (water), and Zeus (air).

During 350 BC, Aristotle made more advancements on the theories formed by Empedocles. He did so by providing explanations on the elements based on their unique qualities and traits.

For instance, you can classify water and fire as opposites, because the former characterizes cold and wet qualities while the latter characterizes hot and dry. The earth and air elements also have the same scenario, with them having dry and cold qualities and wet and hot qualities, respectively.

Based on that point-of-view, dry and wet are the main or primary traits of the elements. The wet trait is associated with flexibility and fluidity, because it seems to be capable of letting things adapt naturally to outward or external situations and conditions. With that knowledge, it is no longer surprising to see wet things being expansive and volatile. You will notice them having the ability to fill all the spaces no matter what their surroundings are.

Dryness, on the other hand, can be identified as a trait linked to rigidity, as it allows things to define their own unique shapes and bounds. You will notice that elements under the dry category are structured and fixed. The reason is that they are naturally capable of defining their own structure and form.

With all these concepts and theories in mind, Aristotle predicted and forecasted the ability of materials to be transformed or transmuted into other things. This is made possible just by changing the combination of the material's qualities and archetypal elements.

The Four Elements and the Symbols That Represent Them

The four elements of alchemy can be identified using the symbols that alchemists in the past and present use. These symbols indicate where each element originated as far as archetypes are concerned.

Here they are together with the actual things represented by each element.

The Water Element

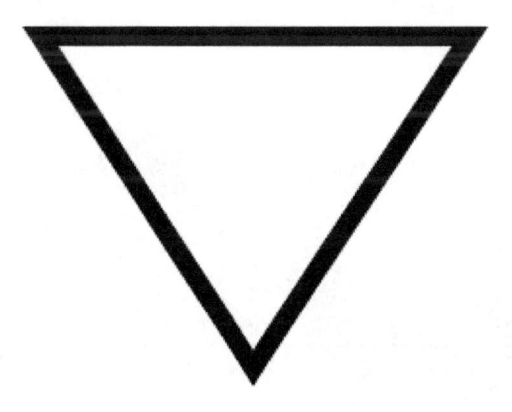

The water element is characterized by the symbol of a triangle pointing downward. The downward triangle is a representation of women and femininity.

This element signifies the calm (phlegmatic) humor of phlegm. In other words, it symbolizes the clear body fluids that your lymphatic system carries and then secretes via your mucous membranes.

A person characterized as phlegmatic is someone who has the qualities of both wet and cold based on the perspectives of Aristotle. This is what makes someone capable of staying in touch with their feelings, though there are instances when they may get broody and moody.

Moreover, you can consider dissolution, union, transformation, and diffusion as forming part of the water element. If you have predominant phlegmatic humor, then it is possible that you are also flexible and flowing. This means that you allow your feelings to be your guide in almost everything. It also makes you focus more on attaining emotional harmony.

Water is also the element of subconscious and emotion, which makes it quite the opposite of air, the element of intellectualism and the conscious. Apart from that, water is an element of physical existence, making it capable of interacting with your physical senses. Though it is less material when compared to earth, it is still superior considering the fact that water holds more activities and motion when compared to earth.

The Earth Element

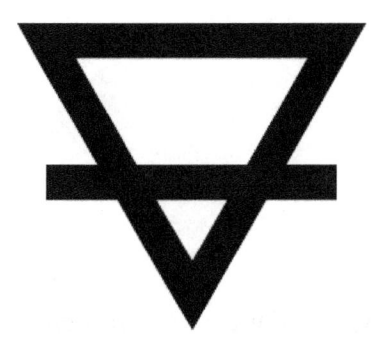

The earth element is the symbol of a triangle pointing downward but is made different from water as you can see; it also displays a horizontal line. One essential quality of the earth is that it is cold. It almost always seeks ways to descend. It has a dry composition, which seems to block the complete descent—the reason why the downward triangle has a horizontal line.

With that said, this element is designed in such a way that it can get suspended in space and time. Alchemy practitioners also believe that it is the most fixed and the least volatile when compared to other elements. According to Plato, the earth symbol has properties that one can connect with dryness and coldness. He also believed that the colors that greatly represent this element are green and brown.

In the field of alchemy, the earth element serves as the symbol for movement and physical sensations. It is an element that indicates stillness, materiality, fertility, stability, being grounded, and potential. Alchemical earth also signifies death and rebirth or beginnings and endings. The main reason behind it is that life originates from the ground and then goes back into the earth to decompose upon death.

The Fire Element

As you can see in the symbol, the fire element simply uses a triangle pointing upward. Despite the simplicity of its symbol, there is nothing simple about this element, as many alchemists believe that it is the most volatile out of the four elements. It points upward as this element always seeks to ascend.

The symbol can be linked to red and orange and is classified as masculine or male. It is also the element which stands for the qualities dry and hot. Traditionally, it was perceived as the most spiritual and rarefied physical element due to the masculine properties it possesses.

Another thing that makes it different is the lack of physical existence. It is capable of producing light and possesses a transformative power, especially when it gets exposed to more physical material. Being the fire element, it is no longer surprising if it is also seen as the representative of fiery and strong emotions, such as love, hate, passion, anger, and compassion.

Moreover, you can easily associate it with life force, blood, strength, and movement. Some also view it as an element that is not only highly protective but also purifying. It is capable of consuming impurities while trying to return them to darkness.

The Air Element

The air element symbol comes in the form of a triangle pointing upward. It also has a horizontal line intersecting at the center. One recognizable quality of the air element is that it is hot. Air also always seeks to ascend. However, it also has moist properties blocking its complete ascent. You will notice that air is suspended in space and time, trapped between those classified as the extremes of below and above.

The air element corresponds to beginnings, creativity, and intelligence. It is hugely immaterial, as it does not have a permanent form. It is active and masculine, making it superior to two of the known material earth and water elements.

Other qualities that the air element represents are optimism and personal integrity. Those who were able to hone this element were also known to be changeable and flighty. Apart from that, alchemical air represents life-giving forces, such as breath. It also signifies the Holy Spirit.

Key Qualities of the Elements

When trying to understand the traditional alchemical elements, you should know that they all have their individual qualities. Every element in alchemy consists of two traits or qualities. It also shares every identified quality with another element. Here are the identified qualities of the elemental system implemented in alchemy:

- **Warm or Cold.** Every alchemical element can either be cold or warm. The quality of the element in this area also corresponds to its gender—meaning whether it is feminine or masculine. The system followed here is pretty basic. You will note that male qualities tend to symbolize activity (motion), warmth, and light, and the female qualities signify passiveness, receptiveness, coldness, and darkness.

    One thing that will help you determine whether an element is warm or cold or if it is masculine or feminine in quality is the triangle's orientation. The masculine warm elements are the ones with triangles pointing upward. This means that they ascend to the spiritual realm. The feminine cold elements, on the other hand, are symbolized by triangles pointing downward, which means that their focus is to descend to the earth.

- **Dry or Moist.** The elements can also be either dry or moist. In this category, you should know that dryness and moistness do not correspond to other elements and concepts straight away. In other words, it is the classification where there are opposing elements. Since every element has one quality being shared with another, one of the elements may also become fully unrelated.

Let us take air as an example. The air element is warm, making it the same as fire. It is also moist, just like water. Despite the two similar qualities, the air element does not have anything to do with the earth. This makes the air and earth two opposing elements.

They are also distinguishable by the lack or presence of a crossbar inside the triangle. Earth and air are opposites with a crossbar in their symbols. The same is true for fire and water as they are also opposites, though their symbols do not contain the crossbar.

The elements also follow a sort of traditional hierarchy based on the qualities they hold. This hierarchy puts the more physical and material elements in the lower part. The elements on the higher level of the hierarchy are also considered as less physical and rarer as well as spiritual.

With that concept in mind, it is no longer surprising to see the earth—being the most material out of the elements—at the lowest. It is followed by water, air, and then fire, which you can see on top of the hierarchy as it is classified as the least material when you compare it with the four alchemical elements.

How to Balance the Elements

The four elements serve as archetypes that you can expect to be present within the collective subconscious. With that said, everyone holds these elements inside of them. In the fields of alchemy and psychology, it is crucial to reach the goal of finding the proper balance. It is what you should be trying to achieve if you want to make the most of the positive outcomes this balance can provide you with.

By ensuring that the four elements are perfectly balanced, meaning they have close to equal proportions within you, you can raise your chances of developing more intelligence compared to others. This kind of balance is also what you need to get the most out of alchemy, especially when it comes to acquiring the most genuine insights about reality.

Apart from that, finding the right balance between the elements of alchemy is also crucial to achieve personal transformation within your psyche. The depth of the relationship between each element, regardless of whether they complement or oppose each other, can also help you to identify whether you will have a balanced and happy life or form psychological disturbances within yourself, like phobias and neuroses.

It is also important to note that all elements came from impressing the cold and hot qualities as well as the dry and moist qualities. You can change one element to another just by changing its qualities. For instance, let's take a look at the cold and moist qualities.

If you impose them, you will most likely get the resulting element in the form of water. You can change this element into the air by boiling it. The reason is that steam forms every time the water boils. When doing that, expect the cold qualities to be replaced by hot. It is also crucial to note that every time a couple of opposing elements meet when they are part of one's personality or because of a certain situation, it can have three potential results: psychic energy generation, neutralization, or union of the elements.

Note that since your goal is to become an alchemist, the result that you should be aiming for would be the third choice—the combination or union of elements, no matter how opposite they are. It can help build transcendence and supreme unity, even on polarities that are otherwise considered as conflicting.

You can also achieve such a result by balancing the elements. Here are the steps that you can take to get such a balance:

- **Identify your predominant element.** Just refer to the descriptions of the elements we provided earlier so you can figure out which one among them tends to govern you the most. When making the selection, it is necessary to practice objectivity. This means spending a bit of time analyzing what each element means and determining the primary element that perfectly suits you.

If you have a hard time identifying your predominant element, you can consult some information on the elements that will guide you through the process. Another way to do it is to have a close family member or friend analyze you. Ask them how they classify you. Their answer will certainly give you a hint about the most predominant trait and element inside of you.

- **Develop the perfect balance of the elements within you.** You can do that with the help of the rotation of the elements used during ancient times. It is a procedure primarily based on the square of opposition developed and introduced by Aristotle. It signifies how exactly the elements and qualities correlate.

The four alchemical elements placed at the corners of a square have qualities that form a diagonal cross signifying opposition, which is the principle to stick to. Any change in the element's quality may result in some motion of the cross through the square. With that in mind, it is safe to assume that the conflict that the opposites have is what serves as the rotation's motor. For instance, hot becomes cold and vice versa. The same is also true for the other qualities—the moist turns to dry and vice versa.

- **Look for the opposite element of your dominant one.** You need to identify the exact element that is dominant in you, as doing so will let you find the opposing one that has sometimes been neglected. Once you find it, your goal is to increase its presence in your life. It is necessary to balance your own temperament, but you have to remember that due to the opposing features of the elements, you also have to handle one adjacent element.

Let's say, for example, water is your dominant element, so you set a goal of balancing it by increasing the level of fire within your persona. What you should do is to start to work on either earth or air—the adjacent elements. Just choose which one between these adjacent elements provides you with greater affinity or comfort.

If your choice is air, then what you have to do is increase the moist qualities. In other words, you have to work on transforming yourself in a way that you start getting more emotional energy to show itself or by making yourself more flowing. If your choice is earth, then the opposite is what you need to do. This means you have to minimize the part of your nature which is flowing and improve your ability to take control of all your emotional energies.

It's not that difficult to balance your alchemical elements. You can even do it through meditation and chanting. You can also achieve this balance by managing your chakras and keeping them balanced.

What is Quintessence?

It is also crucial to get to know more about the quintessence linked to alchemists if you want to gain a full understanding of what the elements mean. Generally, the quintessence is defined by those who practice alchemy as being the fifth element. It has such a reference not because it is part of the natural elements associated with alchemy but because its function and form go beyond the other elements.

Even if this fifth element was only recognized recently, its creation is so wonderful and incredible that it goes beyond the limitations set by the other four alchemical elements. If you are still unfamiliar with the quintessence, then be aware that in alchemy, it is a luminous element, though it is not visible to everyone.

During the medieval period, alchemy viewed quintessence as similar to the elixir. Most people even thought of it as something with a similar magical component or ingredient. It was also believed to be the element that composed the stars. It is the reason why some believed that each living being has a star concealed within, referred to as quintessence. With that said, it is not surprising to see quintessence as holding the symbol of a star.

Quintessence also uses a pentagram as another of its famous symbols. You can see this pentagram being inscribed within a circle and has five equal parts. This pentagram is believed to signify a human's body. Quintessence was also said to be strongly associated with chemicals.

Based on traditional methods of thinking, some chemical components and mixtures are more valuable compared to others. The component or mixture considered as the better one is what is always referred to as quintessence. For instance, the thing that gives the wine its distinctive character is alcohol, so it is that liquid's quintessence.

Moreover, quintessence strongly embodies the philosophical alchemical principles discussed in the first chapter—mercury, sulfur, and salt. Because it has the three mentioned core elements/substances, it is safe to say that it is part of everyone's life force, making it a vital fifth element in the field of alchemy.

Bring the Alchemical Elements into Your Life

The four alchemical elements—earth, water, fire, and air—exist within and around you. All of them are vital to all life forces. These four essential elements form a huge part of who you are. They are also the major components of the universe. You can expect such elements to transcend the physical, allowing them to manifest themselves not only as unique personality traits but also as energetic forces.

With that said, you need to make the elements a major part of your life. Doing so will bring balance. Note that fire depends on water to quench it, while the earth depends on the wind to move. The way your outlook differs from the others and your own way of self-expression could be influenced by having more fire and only less water in your overall personality. You can correct any negative distinction in your personality by balancing the elements similar to nature.

Also, remember that all elements have their own sets of unique qualities. You can't expect the elements to be inherently good or bad. The reason is that each one has its own set of negative and positive qualities. You just have to balance them, so you can finally make them a part of your life.

The good news is that you have several options when it comes to the four elements that play a huge part in your life. Here are some ways to do it:

Element	Strengths	Weaknesses	Ways to Stimulate
Fire	Strength, passion, and courage.	Obsessiveness, jealousy, and anger.	Do regular exercises. Plan for an adventure. Light a candle.
Water	Trustworthiness, peacefulness, and healing.	Inconsistence, indifference, and prone to depression.	Take a relaxing bath. Increase your intake of water. Consume more fruit and vegetables. Put a vessel filled with water on your nightstand or altar.
Air	Excellent communication, joyfulness, imagination, and creativity.	Self-centeredness and dishonesty.	Spend more time outdoors. Add wind chimes at home so you will constantly be reminded of the air's existence. Complete a creative project.
Earth	Being grounded, logical thinking, and hardworking.	Laziness, stubbornness, and being materialistic.	Practice gardening and add plants to your home. Be in touch with nature. Eat healthy and warm foods.

You can also bring elements into your life by wearing an object or symbol of the specific element that you need more of. For instance, if you are working or living in a place that is too competitive or tense for you, then it is possible that the space is full of fire energy. This calls for the need for a balance of water energy. Just add basic objects or symbols representing water so you can mellow out the negativity in the space.

Here are a few ideas on what represents each element:

- **Earth.** The symbol of the element itself; earthy stones, such as agate, malachite, petrified wood, amber, and jasper; wooden and metal objects; brown or green candles; flowers and plants; and the pentacle. Your root chakra also represents the earth, so make a point of balancing it.

- **Water.** The symbol of the element itself; a water vessel; moonstone; blue candles; a symbol of the moon; a chalice or cauldron; coral, sand collars, and shells. Your sacral chakra also symbolizes this element.

- **Air.** The symbol of the element itself; a wand, feather, or fan; dragonflies, fairies, birds, butterflies; smoky quartz and citrine; smoke coming from incense, herbs, and resin. Stimulating and balancing your third eye chakra is also important for the air element.

- **Fire.** The symbol of the element itself; light, matches; fire opal and carnelian; candles, burning incense or herbs, and athame. It is also represented by the solar plexus chakra.

You can place the representations and symbols of each element in various parts of your home to achieve your desired balance and stimulation. You can also make them a part of you by wearing them. You can even put them in your car, on your desk, altar, and in your room. Remember that these four essential elements of alchemy also serve as the foundations of all things as well as the inner and outer parts of you.

With that in mind, it is advisable to make it part of your practices, magic, and rituals. Consider the elements as vital energetic forces that you have to implement to manifest magic and make rituals successful. You just need to have something to represent each element in your sacred space or altar to take full advantage of its powerful balancing force.

Chapter 3: The Planetary Metals

You can also understand alchemy even better by learning more about the planetary metals governing it. Note that seven metals connecting the seven deities and traditional planets constitute alchemy. There are also specific alchemical symbols that you can use to recognize all these planetary metals.

Each element discussed in this chapter is metal, with each metal linked to a celestial object, a body organ, and a day of the week. The reason behind this is that astronomy largely comes from early alchemy.

Every planet was believed to rule over the related metal during the classical era, not only through the position it holds in the sky but also through its closeness or proximity to other planets that greatly affect the properties of the metal. One thing you will most likely notice when studying planetary metals is that they do not include the planets Neptune and Uranus.

The reason for this is that the symbols were formed even before the invention of the telescope. Before that, only those planets that could be seen by the naked eye during that time were believed to exist.

To give you an overview of the seven planetary metals, check out this table:

Metal	Celestial Body	Body Organ	Day of the Week
Lead	Saturn	Spleen	Saturday
Tin	Jupiter	Liver	Thursday
Iron	Mars	Gallbladder	Tuesday
Gold	Sun	Heart	Sunday
Copper	Venus	Kidneys	Friday
Quicksilver / Mercury	Mercury	Lungs	Wednesday
Silver	Moon	Brain	Monday

It is important to understand the planetary elements and the seven planets connected to them.

Lead = Saturn

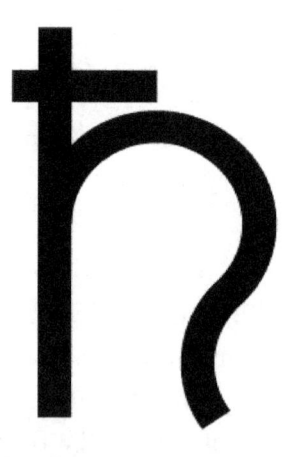

In chemistry, lead is an element in the carbon group. It holds the Pb symbol and the atomic number, 82. It is a malleable and soft metal, which many also regard as a poor and heavy metal. In the alchemical field, lead is famous for being the oldest out of the planetary metals. You can associate it with Saturn.

In alchemical terms, lead comes with an androgynous nature if you associate it with the damp and cold qualities. With that, you will notice some people referring to it as an arcane substance, which symbolizes lusterless prime matter. It has similarities to Adam. The symbol used for it is also like an old man who has a scythe and a wooden leg.

With Saturn as its counterpart and equivalent, you can expect the lead metal to act as either malefic or a purifier. It symbolizes impurities not only in metals but also in humans. Some believe that these impurities refer to troubles and vexations that God placed on His people to encourage them to repent. On the other hand, others believe that these impurities are life's troubles that they have to overcome to reach perfection.

Lead is capable of burning, though. If that happens, then it can also burn all the impurities linked to you. Life's tribulations associated with lead are just temporary because you have the chance to cleanse them. You can cleanse yourself from all the imperfections and impurities you have incurred in your life.

As the governing planet representing lead, the symbol of Saturn is a vital part of the transformation. It can transform you into someone tough and strong-willed. As far as the internal transmutation process is concerned, it signifies your ability to overcome the old Saturn that seems to hold you to fateful determinism.

To overpower such restraints, you can enjoy your newly acquired freedom as you get fully cleansed from impurities. It can make you more capable of working while using your autonomous creativity.

How to Transform Your Life with the Planetary Metal Lead

Incorporate the symbol of Saturn into your meditations and devotions. Doing so can help to bring a higher level of focus capable of facilitating matters that need vigilance for long-term. It is also helpful to promote fluid transitions.

With that, you can go with the flow as you experience changes while keeping your interest detached. It allows you to enjoy the transition without having to take all things personally and let drama take over the situation..

You can also take advantage of those stones known to resonate the energy of Saturn. Among these stones are obsidian, onyx, and jet. What you should do is carry any of these stones whenever you feel like you need the unwavering determination of this planetary metal.

Taking these stones with you anywhere can keep you grounded. Moreover, you will notice their help when it comes to maintaining your stability. Apart from that, these stones will keep you protected while also boosting your confidence. Overall, making this planetary metal a part of your life will let you take full advantage of Saturn's symbolic meanings—among which are domination, methodical, authority, construction, transition, stability, observation, long study, and unrevealed power.

Tin = Jupiter

♃

Tin is also another metal that is so important in alchemy. It is even one of the most important metals used in the alchemical process. It has a symbol that resembles the number "4" with style. Alchemists also refer to the symbol of tin as a crescent beneath a cross.

As a metal, tin is noticeably malleable and ductile. It is a silvery-white and highly crystalline metal with a crystal structure. It serves as a catalyst whenever you use oxygen in a solution. Also, it helps to accelerate the chemical attack.

In the field of elemental alchemy, the symbol signifies the planet Jupiter. With the planet Jupiter ruling it, you can expect tin to have a strong connection with your breath. Some even perceive it philosophically as the natural breath of life. The alchemical symbol of tin also signifies the balance between cold, hot, and mediation.

The symbol also has a close relationship with manifestation, considering how it is situated in the solar system. You can see it resting at the center of the planet's orbital line. You can also see four planets, namely Mercury, Earth, Mars, and Venus, preceding it. It is a position that signifies balance, which is the reason why the Jupiter energy that governs tin is perfect when it comes to finding order, justice, and manifestation.

How to Transform Your Life with the Planetary Metal Tin

Making the planetary metal tin a part of your life can do you a lot of good, especially if you consider the specific areas where its counterpart planet, Jupiter, focuses. These include good luck, abundance, wealth, protection, expansion, balance, justice, supremacy, generosity, and optimism.

Fortunately, it is not that hard to harness the power of both tin and Jupiter. If you intend to incorporate the energy of the planet in order to achieve prosperity and wealth or to improve your confidence and luck, then you merely need to carry stones resonating with the planet's energy. Some stones that you can use are tiger eye, topaz, citrine, and sapphire. You can wear them in the form of accessories or put them in any part of your home as decorations.

Iron = Mars

The third metal we will tackle in this chapter is iron. This metal has an alchemy symbol, which is the male symbol frequently used to signify Mars. It is the reason why many also refer to it as a symbol of masculinity. Being one of the most abundant metals, it is also not surprising to see iron as a metal representing physical strength. The male/masculine energy is predominant in iron.

In philosophical terms, iron signifies the need to control primal urges while embracing the fire inside you. Mars, which is in connection with the metal, is red and this is the reason it is referred to as a fiery planet. This means that it strongly indicates your passion for a lot of things. One more thing that Mars is famous for is its ability to confront issues about life and death.

The connection between the planet Mars and the metal iron also has an incredible influence in terms of illuminating your vision, thereby letting you know the things that are no longer serving or benefiting you. It serves as a force that drives you to start anew, especially after a symbolic death and thoughtful retreat to those who no longer matter.

This planetary metal allows you to be completely honest with yourself because of the strength represented by Mars and iron. You will be more open to yourself, forcing you to confront even your darkest and deepest secrets. Moreover, the masculine energy in this planetary metal will also significantly increase your ability to understand even the things you want to hide.

How to Transform Your Life with the Planetary Metal Iron

To take full advantage of this planetary metal, you just have to honor the wisdom of the planet connected to iron, which is Mars. It would be best to offer this tribute to the red planet during the autumn months. What you should do is to meditate on its energy.

As you say goodbye to summer, while being fully aware of the planet's clarifying power, you will be in for new discoveries. The reason is the power of Mars tends to show what is true even in veiled situations. If you recognize the presence of this power, then you can open yourself up for better self-awareness and expansion.

It would be best for you to try harnessing the power of this planetary metal in every instance you find the need to:

- Boost your confidence.
- Present new ideas to an audience in a way that you can effectively drive your point home.
- Build enough strength to succeed in a heated battle or debate.
- Be passionately inspired to finally begin the project you have in mind.
- Increase your libido by building up your desire.

These are just a few areas where you can maximize the positive effects of the planetary metal, iron, and its counterpart, Mars, into your life.

Gold = Sun

Gold is also another planetary symbol that has a great impact on the alchemical process. It is a prominent symbol in alchemy, which represents perfection. Note that one key goal of a lot of alchemists, though there are many difficulties fulfilling this goal, is turning lead into gold.

One symbol that represents gold in alchemy is a stylized sun that emits rays. In other words, the symbol usually comes in the form of a circle with rays. The symbol also stands for the sun.

An important thing that you have to be aware of about gold is that it is greatly connected to perfection, whether in spiritual, physical, or mental aspects. Most alchemists strongly believe that being able to produce gold indicates the perfection of everything on all levels, such as the mind, soul, and spirit.

It also indicates the goal of most humans to achieve perfection in spirit and mind. In alchemy, you can see people setting a goal of transforming other metals, like lead or iron, into gold.

Doing that strongly indicates your journey as a human to become "golden," which means that you are trying to become better or transform yourself into someone who has a much higher value. As a metal, gold is known for being ductile and extremely malleable. It is non-toxic and often seen being utilized in various fields, including medicine, dentistry, and electronics.

How to Transform Your Life with the Planetary Metal Gold

The planetary symbol of Gold, which is the sun, symbolizes many good things, including life, energy, self, clarity, strength, power, and force. Bearing that in mind, you should do something to make sure that you maximize its effect on your life. You can use it whenever you feel like you need to be successful in your career, get a promotion, improve your health and social status, or approach an authority figure.

This planetary metal is also what you need to harness if you want to be successful every time you speak in public or make presentations. Fortunately, it is not that hard to transform your life with it. You just have to carry with you some elements that symbolize the sun, including gold, bronze, ruby, diamond, and topaz. You can carry any of these elements in the form of accessories or as home decorations.

You may also want to set up a sun altar. Use it to draw the sun's energy and integrate it into your life. Just pick a spot in your living space where you can get a lot of natural light. Put plenty of solar symbols on the altar, too. Examples include gold glitter, a yellow altar cloth, a vase filled with sunflowers, and oranges.

Copper = Venus

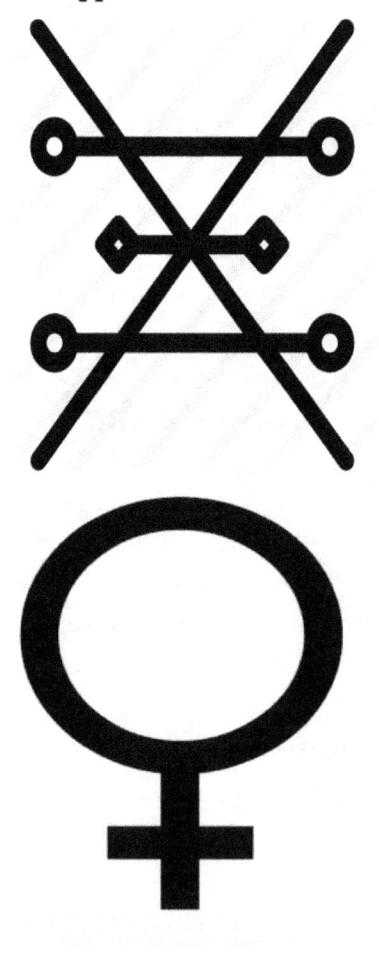

Copper has a couple of symbols representing it—one of which is the female sign, which symbolizes the planet Venus, and the other one is a set composed of horizontal and crossed lines. Most alchemists associate the copper metal with Venus, which is why the "woman's" symbol is used to represent or indicate this element.

One thing you have to be aware of about copper is that it can appear incandescent like the sun. Knowing that, many view it as a solar emblem as it is the only remaining metal that does not have a natural gray or silver shade. It holds a shiny red metal and is considered as being the oldest metal used in a wide range of ancient civilizations.

In the world of alchemy, copper has a symbol that it shares with Venus. You can also associate this symbol with Aphrodite, the goddess, due to the luminescent beauty it possesses. Moreover, copper is a symbol used to refer to harmony, balance, and love. The alchemic symbol of copper also emphasizes the idea of solar power infusion specifically implemented in the field.

You can also see iron representing your aspirations to rise above your earthly and physical desires. This means that the idea behind it resembles a rising sun, which does not achieve perfection, yet. Remember that its elemental alchemy symbol is the same one used for Venus as its planetary symbol, which strongly indicates that this planetary metal embodies many positive characteristics, including balance, love, artistic creativity, and feminine beauty.

How to Transform Your Life with the Planetary Metal Copper

To harness the power of copper and Venus, the planet governing it, you have to make its presence known. For instance, you may incorporate it into your magical wand. Just take the wand you have and then use a copper wire to cover it.

You may also want to use copper for healing purposes. You may have already seen people wearing copper bracelets and saying that these items are helpful in relieving aches and pains. Well, there is nothing wrong with trying to make it a habit to wear these copper bracelets. This is especially true if you want to improve your health and find relief from the aches and pains you often experience.

Another tip is to use those gemstones that strongly represent the planet Venus. You can use either semi-precious or precious Venus-related gemstones or birthstones to get a hold of their healing and spiritual qualities as well as mysterious powers.

One way to do it is to wear a diamond in a silver, gold, or platinum ring. It would be best to do so on Fridays, which is the planetary metal's day of the week. Other gemstones you can use are opal, zircon, and white topaz.

Quicksilver (Mercury) = Mercury

Mercury, a metal that is also referred to as quicksilver, is a central element in alchemy, particularly when practiced by Westerners. It has the exact symbol it used when it was still a vital part of the cosmic womb (Three Primes). Being part of the three primes, you will notice Mercury reflecting a state, which can transcend Earth or death, and an omnipresent life force.

This metal's symbol also has a strong connection to the quickly moving planet, which has the same name. Most alchemists believed mercury could be extracted from all other metals.

They also believed that there was a possibility of producing various metals by changing the quantity and quality of sulfur found in mercury. Gold is considered as the purest of all these elements, and alchemists believed that mercury was necessary to transmute impure or base metals into gold, the primary goal of a lot of alchemists.

Another thing that you should know about Mercury is that its symbol significantly aligns with the planet's symbolic personality. The reason is that Mercury is famous for being the planet highlighting open-mindedness, intelligence, and new and creative ideas. Some even perceive it as the planet of the mind. Apart from that, it also signifies the life force. It has a strong connection with the spirit, which is what you can expect to survive after death.

How to Transform Your Life with the Planetary Metal Quicksilver (Mercury)

Mercury can also help improve your life for the better. The corresponding planet of this metal, Mercury, can harness a lot of positive qualities in your life—among which are agility, versatility, intelligence, perceptiveness, swiftness, fantastic communication, and dexterity.

The good news is that there are a couple of ways to integrate it into your life. One way is to carry stones or rocks that have the energetic reverberations of Mercury. Just put stones, like agate, amber, aventurine, and jasper, inside your pockets to stimulate the positive qualities of Mercury, including better mental focus and self-improvement.

Another effective tip is to grow plants strongly connected to Mercury. Some of these plants are dandelion, which corresponds to bright ideas; lily, which signifies gender equality and pure wisdom; daffodil, which represents cheerful leadership and high-minded communication; and almond, which can stimulate safe travels.

You can grow any of the mentioned plants on the windowsill or in your garden. Doing so works to solidify the concepts of the metal and planet that resonates within each one. Another tip is to keep some dried leaves or roots of the plants inside a sachet or locket and take it with you anywhere you go.

Silver = Moon

The last planetary metal we will talk about is silver, with the moon as its counterpart. It is a symbol that resembles a crescent moon. One can draw the crescent facing either left or right. One thing to note about the metal silver is that the moon it uses for its symbol may not only be the crescent one. It could also be representative of the actual moon.

In the field of alchemy, you will notice silver being a part of the base metals. Most alchemists also use it frequently as a prime material at the starting point of each piece of work. The moon is associated with attaining incredible contemplation and inner wisdom.

You can also associate the planetary symbol with femininity. It has feminine properties. Moreover, it is indicative of intuition.

One more thing about the moon that you have to be aware of is that it has a unique way of welding its influence and force. Even if it is luminary, it still can't produce light on its own accord. It relies greatly on the sun's light to reflect its image to the people of the Earth.

With the manner it uses to project light, it is not surprising to see it symbolizing subtlety. It can also gain reflection, indirect deduction, and full clarity only through passive means. It makes it different from the sun, which is capable of boldly directing its blaze on a philosophical subject. The moon focuses more on softly enfolding your attention and lighting up your psyche using a sheer glow that welcomes more esoteric impressions.

How to Transform Your Life with the Planetary Metal Silver

You can also take advantage of the last planetary metal indicated in this chapter, which is silver, by tapping into the power and energy of the moon. The moon is a symbol of good things, qualities, and elements, including cycles, time, renewal, psyche, intuition, balance, fertility, transition, influence, perception, and illumination.

When it comes to making the most out of the planetary metal silver, the right timing is important. The reason is that you have to take action in a timely manner, and this will fully depend on the different phases of the moon, the planet where silver has a strong connection. Here's a guide to making the most out of these phases:

- **New Moon.** This phase is all about a fresh start, new beginning, or rebirth. It is the moment when the moon and sun come together. This phase is the best time for you to set your intentions, willing them to manifest by letting them reach the universe. One way to do it is to do an ayurvedic cleanse during the new moon of the fall and spring seasons.
- **Waxing Moon.** This phase represents attainment, manifestation, and growth. To take full advantage of this phase, make it a point to schedule the execution of creative projects during this time. It is also the best time to begin writing. It is also the ideal phase for correcting your mistakes and fine-tuning a certain endeavor or project.

- **Waning Moon.** This important phase of the moon usually represents release, contemplation, incubation, quiet time, and surrender. You can let go and surrender by sharing your acquired wisdom and knowledge with other people.

- **Full Moon.** This phase signifies the peak of power, clarity, and fullness. It also means that you are at that stage where you have already obtained your heart's greatest desires. With the many positive things that could happen during this phase, it would be the perfect time for you to formulate an important decision. You may also find it the most auspicious time to launch or start that project you have been putting off.

With the knowledge that you have acquired from these alchemical metals and planets, you have a better chance of unlocking what alchemy is and how you can use it to improve your life.

Chapter 4: The Mundane Elements

Aside from the elements mentioned in the previous chapters, alchemy also has mundane elements. These mundane elements are new additions to the field of alchemy. With that said, each one does not have a long history associated with it. It is for this reason that you can only find limited information about these mundane elements and their representations.

Despite that, it has been proven that several alchemists were able to use these elements successfully at certain points of their practice. Here are the mundane elements that we are talking about.

Antimony

The first mundane element which we are going to talk about is antimony. It is a chemical element. As a metalloid, it has similarities to a metal in physical properties and overall look.

Antimony has a symbol that comes in the form of a circle that has a cross on top of it. It is also characterized by another symbol, which is like an upside-down female sign. There are also instances when a wolf represents this element.

As for its meaning, take note that antimony is an elemental alchemy symbol signifying the animal tendencies naturally present in humans. It means that it can show your natural wildness. It is useful if you want to bring out your strong and free-spirited side.

To use it, you can just wear anything with this symbol, like an accessory. Wear it whenever you feel meek and want to embrace your animal power. You can also see antimony being used in the manufacture of ceramics, glass, and plastic. You may want to integrate the items based on it in your home décor, too, if you want to bring out your animal power and take advantage of it.

Arsenic

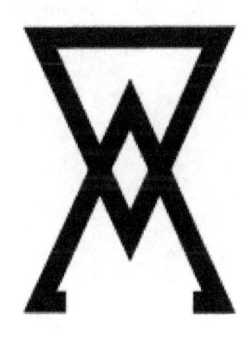

Arsenic is also another chemical element, which has a role in alchemy. If you are unfamiliar with this element, then take note that one important thing about it is that it is a poisonous metalloid with three allotropic forms—gray, black, and yellow. This element and its compounds are commonly used to create insecticides, herbicides, pesticides, and different alloys.

It also has an elemental alchemy symbol, which is useful in creating magic and medicinal cures.

In the field of alchemy, a swan or group of swans often represent arsenic. The reason is that this element has the ability to transform its appearance, which is a yellow crystalline or metallic gray solid, resembling the manner through which a cygnet turns into a swan. You may also see a symbol in the form of two overlapping triangles.

As a metal, the primary use of arsenic is to strengthen the alloys of copper as well as lead. It was also popular among the early alchemists as it formed part of their alchemical rituals that require the use of medicinal and magical cures.

Bismuth

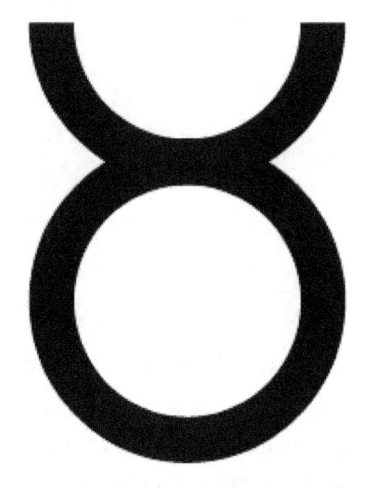

Bismuth is also another mundane element that you have to be aware of. While there is only a little information regarding the way it was used in the field of alchemy, people in the past often confused it with lead and tin. It was only during the 18th century when it somehow got its own personality. It has a symbol that you can liken to the number "8," but the only difference is that the topmost part is open. You may also view this symbol as a full circle with a semicircle on top of it.

Bismuth is distinctive because it is a brittle and crystalline metal, which has a silvery-white color. As a chemical element, you will notice that it occurs naturally, though the oxide and sulfide present in it can develop essential commercial ores.

Bismuth is also a metal famous even during ancient times, though people confused it with tin and lead at that time. It could be because of the physical properties that it shares with the two metals. Going back to alchemy, many consider it as a useful element capable of stimulating energy and vitality. You can use this symbol to achieve your goals, especially if you are working in a team.

The compounds of this element compose around half of its production. You can also use it in things like pharmaceuticals and pigments.

Magnesium

The next mundane element we have to talk about that has an influence in alchemy is magnesium. It refers to a lightweight yet fairly strong metal with a silvery-white shade. It tends to tarnish a bit when you expose it to air. Magnesium is different in the sense that it is quite hard to ignite in bulk. You can easily light it up, though, if you shave it and form thin strips out of it.

You may have a hard time extinguishing it once you have lit it. This level of difficulty when extinguishing it is one reason why many practitioners find this symbol of elemental alchemy appealing. According to them, this symbol is representative of infinite flame, ascension, and eternity.

Many different symbols emerged for magnesium from the moment it was introduced in the field, but the most commonly used one is an unfinished circle with a vertical line dissecting it. You will also notice a horizontal line emanating from the vertical line's median point. Moreover, there's a smaller vertical line, which you can see intersecting the right side.

One thing you should know about magnesium is that it is not available in pure form, which is why alchemists conducted their experiments using magnesium alba or magnesium carbonate instead.

Apart from playing a huge role in alchemy, magnesium also helps create products that need to be lightweight, like car seats, power tools, cameras, laptops, and luggage. Moreover, you can add it into molten steel or iron to remove sulfur.

Other uses of magnesium include making heat-resistant bricks specifically designed for furnaces and fireplaces, and acting as an important component of fertilizers and cattle feed. It also has medicinal uses, particularly the magnesium hydroxide and sulfate forms.

Phosphorus

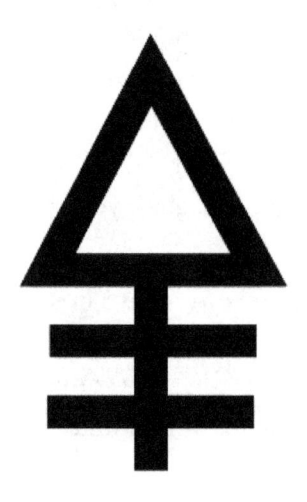

Phosphorus is also another mundane element, which fascinated a lot of alchemists in the past and is still used today. One fun fact about phosphorus is that it was the name used for Venus during ancient times, especially when you see the planet before sunrise.

The most common phosphorus tends to create a waxy white solid with a distinguishing disagreeable smell. The purest forms of it, on the other hand, are known for being transparent and colorless. Generally, it is non-soluble when exposed to water, but you can expect it to be soluble when used along with carbon disulfide. The pure form of phosphorus is also capable of spontaneously igniting when exposed to air and burning to form phosphorus pentoxide.

One thing that makes phosphorus so interesting to alchemists is that it is capable of containing or trapping light. Proof of this ability is its compounds' glow-in-the-dark phosphorescence. The fact that it can capture light is also one reason why alchemists use it to signify spiritual illumination.

Its symbol is in the form of a triangle visible on top of a dual-cross. This specific symbol makes it representative of the spirit, too. Most alchemists also seem to be fascinated by the properties of phosphorus, which made it capable of being burned in the air. Moreover, alchemical tradition strongly associates phosphorus with the fire element.

Platinum

The next mundane alchemical element that we will talk about is platinum. It is a silvery-white and beautiful metal when in its purest form. It is also a ductile and malleable metal with the ability to resist corrosion. The alchemical symbol is a combination of the sun's circular symbol and the moon's crescent symbol.

It earned such a symbol as most alchemists viewed platinum as an element combining gold, which corresponds to the sun, and silver, which corresponds to the moon. One way through which platinum is used in philosophical elemental alchemy is as a representation of endurance.

You can even use it to symbolize your grit and determination. It is also the element you need to see your manifestations coming to fruition or completion. Also, platinum can be viewed as an element that represents superior achievement and a reward for crafting the best version of yourself. It is in alignment with your whole chakra system.

Platinum is also useful in opening yourself up, so you can be more aware of your inner knowledge and peace. As well as being used in the field of alchemy, platinum also plays a vital role in making dentistry and laboratory equipment, jewelry, and thermometers. You can also combine it with cobalt if you want to form strong and permanent magnets.

Sulfur

There is also a mundane element called sulfur. It is a non-metal that has the qualities of being odorless, tasteless, and abundant. When viewed in its natural and traditional form, you will notice sulfur appearing as a yellow crystalline solid. Naturally, you can view it as a pure element or mineral of sulfate and sulfide.

As a chemical, sulfur is capable of reacting either as a reducing agent or as an oxidant. It is capable of oxidizing the majority of metals, as well as a few non-metals, like carbon. With that, you can expect it to result in a negative charge in the majority of organosulfur compounds. It also has a strong ability to lessen a few strong oxidants, like fluorine and oxygen.

You can also produce elemental sulfur crystals out of it, which many mineral collectors seek today because of their brightly colored and unique polyhedron shapes. With its abundant properties, the native sulfur was famous during ancient times, as proven by its uses being recognized and mentioned in ancient Greece, India, Egypt, and China.

One use of it during that time was as a fumigant brought on with the fumes produced after you burn sulfur. There were also medicinal mixtures containing sulfur used by many people as balms and anti-parasitic solutions. In fact, some are still used today.

Another great thing about sulfur is that it is a vital life element. You can find it in a lot of amino acids. You can view it as a transcendent elemental symbol in the field of alchemy, which signifies human nature's multiplicity. It also signifies one's eternal aspiration and desire to achieve enlightenment.

As you can see in its symbol, this element represents the Holy Trinity (the triad of ascension). Sulfur is part of the heavenly substances, together with mercury, and salt, which compose alchemical science. You can also take advantage of sulfur when it comes to getting rid of your negative thoughts, emotions, and anything that tends to block or hinder your progress.

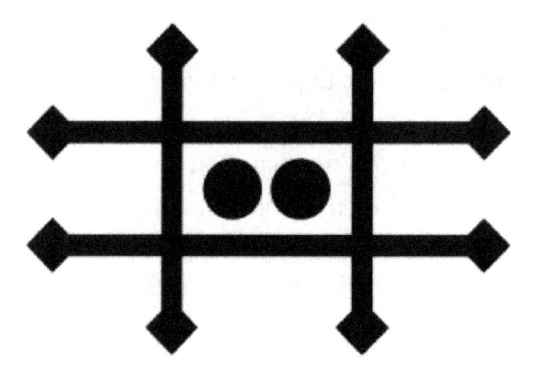

Zinc

This chapter's last mundane element is zinc, symbolized by a matrix that has two sets of parallel vertical and horizontal lines intersecting one another. You can also find tiny diamond shapes attached to the end of each line. Apart from that, there is a rectangle in the center of the lines that features a couple of tiny circles.

As a metal, zinc is recognized for being moderately reactive. You can mix it with oxygen as well as other non-metals. It tends to emit a reaction when combined with diluted acids to release hydrogen.

Zinc also left a mark in the world of alchemy, as practitioners often call it the "philosopher's wool." This wool is produced after burning zinc in the air. Zinc is also useful in boosting fertility, and for property rituals. You can also use it to lessen your grief in case of a loss. Dreams featuring zinc were also recognized as symbolizing self-authority.

Chapter 5: The Twelve Operations 1 – 6

Generally, alchemy is perceived as a process where the ultimate goal is personal growth. It is all about purifying yourself. You are already familiar with the concept of transforming lead to gold. Most alchemists certainly viewed this as their ultimate goal. Several alchemists also viewed the physical process as a reflection of and metaphor for their inner transformation.

This means that the goal of alchemy is to examine yourself and adjust any negative habits that you have, so you can transform yourself into a better human being. The point is to strip yourself of the bad and let your inner beauty come out. It is where you will find the principles, elements, and metals in alchemy mentioned in the previous chapters of this book helpful.

One more thing that you have to be aware of is that your journey toward attaining personal growth and positive inner transformation requires you to go through a dozen alchemical operations. Let us talk about the first six alchemical operations you have to go through to reach your ultimate goal.

Calcination

Correspondences:

- Element - Fire.
- Color - Purple-red and magenta.
- Metal - Lead.
- Planet - Saturn.

Calcination is the first out of the major alchemical operations. It basically means purification using fire. Chemically, the entire process or operation consists of heating a substance on top of an open flame or in a container until it transforms into ash. Sulfuric acid is the substance representing calcination.

Alchemists made this from vitriol, which is a naturally occurring substance. One important fact about sulfuric acid is that it serves as a powerful corrosive capable of eating away the flesh. Sulfuric acid is capable of reacting to all kinds of metals, with the exception of gold.

When viewed psychologically, calcination is all about destructing your ego and your strong attachments to all material things that you possess. With that, it is safe to say that this stage is naturally humbling, since it requires you to assault and overcome your life's tribulations and trials slowly. Note, though, that this operation may also require you to surrender your innate hubris, specifically those gained from various spiritual disciplines, deliberately.

Physiologically, you will most likely experience calcination in the form of aerobic activities or metabolic disciplines, and these will be capable of "tuning" your body to make it stronger. It helps you to get rid of all the excesses you have in your life, especially those caused by overindulgence. Similar to meditation, it allows you to discover your inner self.

When that happens, you can naturally lessen your bodily awareness, making it possible for you to raise your consciousness toward more subtle energies—the ones that serve as your life foundations. In other words, it helps you in letting go of all unimportant drains on your energy, and leaves only those that play significant roles in your life.

Coagulation

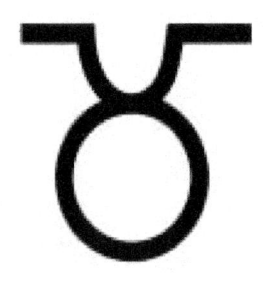

Correspondences:

- Substance – Salt.
- Color – Purple and violet.
- Metal – Gold.
- Planet – Sun.

Coagulation is also another important operation in the field of alchemy. It means the inseparable unification or union of volatile substances, fixed into a single mass. This mass is extremely fixed in the sense that it is capable of withstanding even the most violent and the harshest fires. It is also capable of communicating its fixed nature to transformed metals.

In the psychological sense, this alchemical operation can be perceived as your newly developed confidence that tends to go beyond everything that is around you. Some of those who reach this phase also experience coagulation in the form of a second body—consisting of a golden coalesced light. It serves as their permanent vehicle of awareness, embodying supreme aspirations as well as mental evolution.

Through this alchemical operation, you can also incarnate and release your soul's "Ultima Materia," characterized by the astral body. Alchemists also call it the philosopher's or the greater stone. The use of such a magical stone is the key toward alchemists strengthening their belief regarding their ability to exist on every level or phase of reality.

Similar to the dissolving phase, coagulation can also be considered as a cyclic process. You can often take it in terms of the volatile's final fixation. It means refining the mind to a certain extent, turning your divine essence's awareness into something permanent. This makes it so fixed and permanent that nothing can throw it off.

In a physiological sense, it is the alchemical operation signified by the elixir's release into your blood. It helps to rejuvenate your body, restoring it so it can reach its highest peak of health. The psychological and physiological processes involved in this phase work in creating your second body, composed of solid light, which tends to come out through your gold or crown chakra.

Fixation

In this alchemical operation, the fixed state is what you will be trying to achieve because it is the process or act of attaining firmness by stopping fluidity. The fixation process involves transforming what was a volatile substance before into another form that even a strong fire can't affect. This new form is usually solid. It works by separating an object or substance then bringing it back in either a different or similar shape or form. This transformation also happens on a sub-atomic level.

In alchemy, fixation plays a major role, considering that this is the process needed for the successful transformation of an object or substance. It is also required to complete the magnum opus in the field of alchemy.

One more fact about the fixation alchemical operation that you have to be aware of is that it does the fixation of the volatile continuously. In other words, it is like a never-ending process starting from blackness through whiteness. A sign that this stage already reached the highest level is when it gets into the redness fixation.

This alchemical operation is also almost the same as coagulation. The two are also different in the sense that fixation is often taken as an ongoing procedure of dissolving and fixating. It continues from the start to the end. With that, it requires you to increase your awareness of the divine essence and your inner mind and ensuring that it forms part of your daily life.

Dissolution

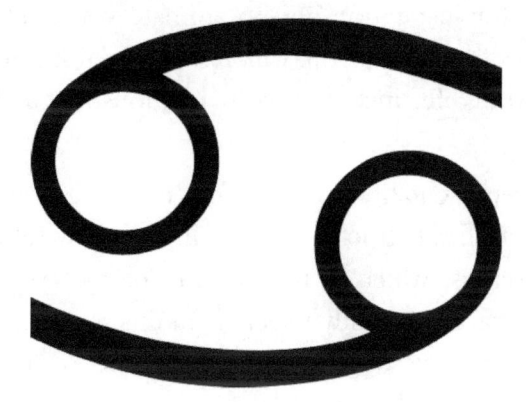

Correspondences:

- Element - Water.
- Color - Light blue.
- Metal - Tin.
- Planet - Jupiter.

Dissolution is the alchemical operation characterized by reducing the body in such a way that you turn it into its elemental principles of primal matter. In a chemical sense, the entire process involves the use of water to dissolve the ashes derived from the calcination of the alchemical operation. Rust or iron oxide represents the phase of dissolution. This strongly indicates water's corrosive powers when it comes into contact with even the toughest metals.

Dissolution is also a vital aspect or stage in alchemical transformation, since it plays a vital role in making you fully aware of every existing life. Here, you will be required to blend with water, which is considered to be the solvent containing and supporting you and every other human life. Water is the universal solvent used in this operation. It is unity, grace, love, and connection.

By blending yourself into the water, you can loosen your boundaries without locking your body and yourself into your identity. You float and go with the flow. It also allows you to strengthen your devotion so you can use it for inspiration and to animate your daily actions. With your stronger devotion, you will have something to influence every aspect of your life, including your relationships and career, positively.

Dissolution is also a representation of yourself breaking even further from your psyche's artificial structure. You can do so by fully immersing in your unconscious, which is the specific part of your consciousness that you may have constantly rejected. In this stage, you will notice this operation's dissolving water come to you in the form of visions, strange feelings, voices, and dreams.

When that happens, you can view them as revelations that there is an existing world that is less rational and organized out there, one that tends to have an effect on your daily life. By letting this operational alchemy come naturally, you give your conscious mind the chance to let go of the need for control. It is a huge help in letting buried materials, as well as your tied-up energy, come out.

You will know that you have successfully experienced dissolution if you experience the bliss and satisfaction of going with the flow. This means that you experience genuine happiness from using yourself in the right way and actively taking part in creative activities while ensuring that an established hierarchy or any personal hang-up does not get in the way. If you want to take full advantage of this alchemical operation and grow steadily using it, then stick to a nature-based, agrarian, and monastic lifestyle as this can help.

Digestion

The next alchemical operation highlighted in this book is digestion. Alchemists consider digestion as the fifth vital step in the process of alchemy, where the philosopher's stone is being confected. This operation can also be attributed to the zodiac sign Leo. This zodiac governs the spinal column and heart in the human body.

However, take note that the Kundalini, which refers to the serpent power and is found within the spinal column—yet is rooted in the chakra known as Saturn—is the one that will be redirected through the process eventually. With that, expect your heart center, which is considered as the gold, to be opened.

Remember, though, that you can't expect that to happen and unfold if you do not prepare your physical body completely for it. You have to prepare the cellular structure of your body sufficiently so it can withstand the extreme invasion of light-power.

One more thing that you should know about digestion is that it involves the process of transforming one substance to another. The resulting product should be something that you can use. The energies you can find and access in the inner world are the results of such transformations.

Digestion is also a step that you should avoid viewing as a mere phase of operation in the entire alchemical process. Remember that properly digesting healthy and sensible foods is essential for you to live a happy life, even if you are someone who has no interest in alchemy. This means that digestion is something that you have to familiarize yourself with, considering its importance in many aspects of your life.

Speaking of digestion as it relates to food, it is crucial to remind yourself how important it is to start the process in the brain. Note that your digestive juices begin to flow every time you smell and see food. It is the reason why most foods can cause your mouth to water.

Taking more time to appreciate the food in front of you and savoring its scent before tasting it can further stimulate your digestive system.

Distillation

Correspondences:
- Substance – Mercury.
- Color – Deep blue.
- Metal – Silver.
- Planet – Mercury.

Distillation is the last alchemical operation we are going to tackle in this chapter. The remaining ones will be discussed in the next chapter of this book. In a chemical sense, distillation refers to the process of increasing the purity of a fermented solution through boiling and condensation. One example is when you are trying to produce brandy out of distilling wine.

In commercial areas, distillation has several applications—one of which is separating crude oil to produce several fractions of it for a wide range of uses, like heating, power generation, and transport. Another application of distillation is the removal of impurities from water (for example, getting rid of salt from seawater). It is also possible to distill air as a means of separating its parts, particularly argon, oxygen, and nitrogen, so you can use it for industrial applications.

Distillation is a crucial process for each level of alchemy. It is essential in the sense that it is the process designated to release volatile essences from matter, turning them into their purified forms through condensation. It is also all about agitating psychic forces to remove impurities from an inflated ego. Therefore, it has a goal of completely purifying you, so you can go smoothly to the final stages of alchemy.

If you implement personal distillation, then be aware that it consists of various introspective methods that aid in raising your psyche's content to the supreme. It makes it devoid of any emotion and sentimentality. You have to cut all these off, especially the unwanted emotions from your personal identity. Distillation is also all about purifying yourself, specifically the "unborn" self. It is therefore the key to being what you truly are and becoming who you want to be.

If you look at it on a planetary level, you can consider distillation as the process of realizing and discovering the real power of supreme love. It allows you to reach such a realization because the whole planet's life force continuously seeks to be one with nature's forces, depending on the truth of a shared vision.

Chapter 6: The Twelve Operations 7 - 12

Now that we have discussed the first six alchemical operations, it is time to delve deeper into the subject by understanding the remaining phases or operations in the field. Here, you will learn more about sublimation, separation, incineration, fermentation, multiplication, and projection—all of which are the specific operations you need to go through to reach your desired transformation.

Sublimation

Sublimation is another alchemical procedure or operation you have to be aware of if you want to get more familiar with this subject. This alchemical operation is all about the dissolution and reduction of matter to its principles so it can achieve complete purification. The kind of purification brought on by sublimation seems to help in making all heterogenous and terrestrial components subtler.

With that, you can expect them to receive the kind of perfection deprived of them. You may also choose to let go of the chains imprisoning the mentioned components and preventing their potential growth. Note that your daily consciousness is usually limited only to your conditioned instincts, reflexes, and programs.

Your consciousness is also sometimes extremely immature, so it may be necessary for you to sublimate it, dissolving it to its fundamental energies, and purifying it completely. Once completely purified, you will most likely finally achieve perfection. Keep in mind that in essence, you are already perfect, but all your impurities seem to be obscuring or blocking this perfection. You can bring such perfection back by undertaking the state of sublimation.

What is great about starting sublimation is that it allows genuine and ecstatic love to come in. It gives you the chance to experience an unconditional and all-embracing kind of love—one that is unbound or unattached to anything. The kind of love you will feel in sublimation is not necessarily brotherly nor romantic. It is just love—the genuine and pure kind.

Also, note that in sublimation, actions tend to flow not only from love but also from wisdom. It has such powerful purifying properties that even chemists use this technique in purifying compounds. They do so by putting a solid in an apparatus specifically designed for sublimation. They then heat it beneath a vacuum. Under the decreased pressure, you will notice the solid volatilizing and condensing to form a highly purified compound. This means that it has already left behind any non-volatile remains of impurities.

After that, they collect it from the surface where they have let it cool down. Anyone who intends to increase the purification efficiency of this technique even further can also put on a temperature gradient. It makes it easier to separate various fractions.

With this stage focusing more on purification, expect sublimation to make you enjoy a new experience linked to reality and self—both of which have redefined versions.

The good news is that you can still adapt to cultural and social situations even without succumbing to any identity, especially an unwanted one. Sublimation, therefore, does not oblige you to avoid nor attach yourself. You just have to go through a simple fluid motion when going through your life.

To further expand the possibilities brought on by going through sublimation and practicing it, you can try to explore through these avenues:

- **Connect with nature.** You also have to commune with the creatures that inhabit nature. It allows you to learn a lot, including better perception and improved ability to communicate and connect with others. This tip involves learning more about plants and animals and the minerals offered by the natural world. It also allows you to keep in touch with nature spirits and elemental beings.

- **Transmit energy to areas needing it the most.** For instance, you can send your loving energy to a group and a corporate or political leader. By sending this energy, you can inspire shifts that can greatly benefit the collective.

- **Offer support to those who need it.** One thing that you can do, in this case, is to offer your direct support, especially to those who are still understanding the ways for them to navigate their unique awareness.

Separation

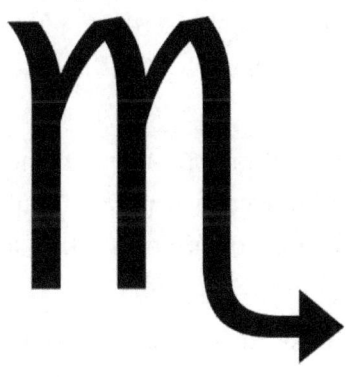

Correspondences:

- Element - Air.
- Planet - Mars.
- Metal - Iron.
- Color - Orange-red.

There is also what we call separation. It is another crucial phase or operation in the field of alchemy. Separation is usually the result of the body's dissolution through its solvent. Such separation takes place whenever the matter turns to black—after which the elements' separation starts.

The black color will transform into vapor, which signifies the earth also turning into water. Such formed water can be expected to condense then go back into the earth. In that case, it will restore its white shade, which usually resembles the air's whiteness. After the whiteness, expect redness to come next—it can signify the air turning into fire.

With all the changes that can happen in this phase, you can view separation as quite similar to the body's dissolution and the spirit's coagulation. It is all about completing the entire process of alchemy through powerful means by uncovering the strong desire that motivates you to reach your goal.

Note that if you do not uncover this strong force of desire that motivates you, it is highly likely that you will get lost during the alchemical transformation process. It would be hard for you to find your way out if you do not have a goal.

Fortunately, you will enjoy several positive things by keeping the specific parts of your life that are usually unnoticeably separated and distinguished. Remember that as a human being, you are most likely a meaning-making machine.

This means that it is natural for you to make up stories, like about how you think others view you. In other words, there are several times when you are not truly present. This may make it harder for you to appreciate what is truly in front of you.

By going through the separation alchemical phase, you can avoid such issues. It even allows you to move the otherwise subconscious drives through your consciousness, so you can scrutinize and analyze each one. By doing that, you can choose how to build relationships with them instead of letting them drive you.

One way to make the most of separation is meditation. Here, you can separate your awareness from your own bodily consciousness. Once you meditate, your first experience after closing your eyes will be blackness. Despite this blackness, meditation will still help you discover your inner world from the time you start the separation phase.

One thing you may discover is that the energies in your inner world can be likened to water if you compare it to the earth or body. With that, you can choose to refine such an inner world. The good thing about doing this is that it can help turn your experience into a subtler one.

Another great thing about reaching the separation alchemical operation is that you get the chance to separate some parts of a whole, especially the ones you haven't noticed in the past. By doing that, you can finally look at all your relationships and see things about them that you were unable to see.

It also lets you contrast and compare various parts of yourself that compose your humanity, specifically those that you find so rewarding. With that, you can make your experience more colorful. It can also integrate rhythm to each step that you take. Just make sure to exercise integrity and honesty when evaluating yourself and acknowledging the specific parts of you that are not acceptable. Once that happens, you can bring yourself closer to your goal of attaining wholeness and perfection.

Incineration

The next alchemical operation you have to learn is incineration. This can be defined as an action involving adding more mercury to a particular matter that is turning into sulfur. The act is done to multiply the result or output or create the perfect tincture or elixir.

Many alchemists consider incineration as the 9th step or operation necessary in manufacturing the philosopher's stone. You can see this specific placement being listed in Manly P. Hall's *Secret Teachings of All Ages*.

Being one of the most crucial steps in the process, you can also attribute it to Sagittarius, which is also the 9th zodiac sign. Also, since the planet Jupiter is the one that rules the zodiac Sagittarius, you can expect it to be the key or chief power of the alchemical operation incineration.

The previous steps and operations in the alchemical process often focus on stimulating your lower self to purge separative and erroneous habits that form part of your personal subconscious. However, you have to note that those are not the only ones you have to remove.

Keep in mind that you can also find several separative and negative elements deep within you that you can reach from a conscious level. Such elements are those that you have to burn from within while you are also working on your outer elements. This is the process that the stage of incineration wants to focus on. In this stage, you will have to use your higher self in crowning your endeavors.

Incineration can help you with that as it focuses on metaphorically burning your brain clean from any life impressions in the past that are still inhibiting you. This operation also works in burning the impressions derived from race consciousness. The best way to succeed and perform this process at a conscious level is through meditation. You have to meditate so you can work on finally removing all the unwanted elements from your consciousness.

It is also crucial to combine meditation with a complete understanding and a strong belief in everything that transpires. Moreover, you need to pray intently for guidance so you can turn over each detail of your life to the higher genius. You can do that while having complete and absolute faith that you will soon realize your intended union with One-Life.

Fermentation

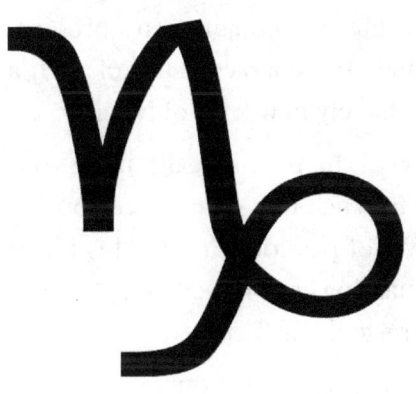

Correspondences:

- Substance – Sulfur.
- Planet – Venus.
- Metal – Mercury.
- Color – Turquoise.

Now, let's talk about fermentation. This crucial alchemical operation is a process composed of two vital steps—the first one being called putrefaction. This first step allows the matter to rot and decompose first before fermenting it or making it live again in spirit. The putrefaction phase associated with the alchemical operation fermentation has images linked to graves, rotting flesh, coffins, corpses, funerals, mutilations, guardian angels, worms, and other related images.

Once this phase is completed, fermentation comes, which is often characterized by refreshing and positive images, including scenes of germination and sowing, rebirth, and greenness. There are even instances when fermentation shows a king and queen with wings. They may also appear as angels. These representations aim to emphasize the spiritualized form of the new being.

If you look at the two-stepped process of fermentation, you will realize that it aims to introduce a new life to a conjunction's product. This phase's main objective is to change the characteristics of such a product completely, raising it to a completely new level of being.

In this phase, you will have to let go of the earthly realm using your imagination. Your end goal would be to reach that state where you can set your soul on fire with a higher level of passion. You let this phase start by inspiring or rousing the spiritual power coming from above. It aims to reanimate, enlighten, and energize the alchemist.

Physiologically, the fermentation stage requires you to rouse the living energy, otherwise known as kundalini or chi, within your body. You have to do that so your body can start healing and vivifying. You can also express it through spoken truths and vibratory tones that emerge from your mercury or throat chakra.

The good thing about fermentation is that its beautiful results are achievable through various means and activities—among which are intense prayer, transpersonal therapy, deep meditation, and personality breakdown. You may also want to do the following:

- **Honor and respect all feelings and emotions.** You have to do this without allowing yourself to get too attached to them.

- **Maintain a wide space for humility and curiosity.** Note that there are still many things that have to be revealed. In this stage, you will be in an unending path or process with every perspective being relative, so you should always have a space within you that should fill up your curiosity.

Overall, fermentation serves as a living inspiration that there is always something that is completely beyond you. The entire experience of fermentation even serves as the foundation of mystical awareness and religion up to the present. It is a truly crucial process throughout your journey toward attaining alchemical transformation.

Multiplication

Multiplication is another operation that embodies how an alchemist masters their craft. It involves the process or act of increasing or multiplying the volume or quantity of something.

The good thing about the multiplication operation is that it allows a significantly big, multiplied increase in the redemptive effects of anything that an alchemist creates. Multiplication also involves doing a completed operation again but with the goal of perfecting or exalting the effects of the substances.

Based on what a certain philosopher believes in, the secret to attaining the best results out of multiplication is to encourage a physical reduction of the object into mercury. It also aims to reduce its primal matter. In other words, you can get a hold of the matter prepared and cooked by nature then reduce it in such a way that it goes back to its first or initial philosophical mercury or matter where it was derived.

In most cases, this stage takes place when the Magnus opus is almost at the end. This is so it can raise the gains in the next projections. Apart from that, multiplication applies to the alchemical goal of reproducing gold and silver.

Projection

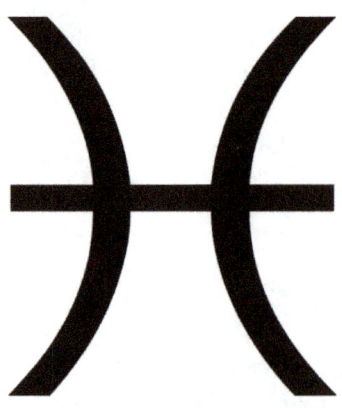

The last alchemical operation is projection. Many consider it as the ultimate objective of Western alchemy. After the successful creation of the powder of projection or the philosopher's stone, you can use the projection phase in transmuting a substance of lesser form into a much higher one. The usual example is transmuting lead to gold.

In most cases, the projection process involves casting a tiny part of a stone to turn it into a base metal in molten form. Another fact about projection is that the entire metal where you project the powder can't be fully transmuted to either gold or silver in case the powder remains unpurified before throwing it into the mix.

Projection is also one of the operations used in completing the coagulation process. It serves as an extension of your emotional energy and thoughts to the world. The result of projection is genuinely good for everyone. However, when taking part in this process, you have to be extra careful as its early stages can be quite harmful to you.

Let us use anger as an example. When you feel angry or negatively overwhelmed by the things in front of you, it is highly likely that you will harbor the energy of a similar negative quality within you subconsciously. Yes, you can let yourself express such negative feelings and energy externally, but there is a high chance that the energy triggering your anger is concealed from yourself.

If you leave it uncontrolled, then such negative psychic energies may build up. The buildup may be so tremendous that there is a chance for them to burst anytime. It is why you have to be extra careful once you are already at the stage of projection.

Your goal should be to get the actual result of alchemical projection so you can take advantage of its power, especially when it comes to attaining perfect equilibrium and health inside. Such a result can be expected to manifest as pure external and internal beauty.

Chapter 7: Herbal Alchemy

Herbal or plant alchemy refers to a powerful in-depth and evolutionary process, which involves purifying a plant and making it evolve so it can reach its highest form. It is an ancient art, which gives you the chance to create a plant or herbal remedy with the aid of alchemical principles.

During this process, you will take advantage of alchemy, which utilizes naturally occurring processes designed to promote evolution and transformation not only to substances but also to yourself. You can also view herbal or plant alchemy as a mystical variation of herbal medicine, created with alchemy and astrology in mind.

It is a fantastic way of deeply learning and understanding a plant, which would otherwise require an extremely long period (even decades) of experience to understand.

With that said, it is safe to say that plant or herbal alchemy is not only a physical process. You can't expect to learn and understand it by just using pictures and words. This means that you have to get a full experience of it so you can grasp what it is truly all about.

Many plants and herbs were already recognized throughout history for holding and providing a lot of powerful benefits in various areas. All you have to do to take full advantage of these plants and herbs is to ingest, brew, burn, or wear them.

Herbal or plant alchemy takes advantage of such power. This field uses specific herbs and plants because of their enhancing properties. Most of these herbs and plants also have the power to let a ritual focus on a certain outcome, preferably that which the alchemists desire. With that in mind, herbs and plants are truly valuable in the process.

Planetary Metals in Plants and Herbs

One reason why plants and herbs play a major role in alchemy is that there are planetary metals in them that you can harvest and use. Note that just like humans, there is also a strong planetary energy ruling plants and herbs. Similar to metals and minerals, plants also have their own planetary correspondences. With the help of this planetary correspondence, you can determine the spiritual and physical effects of each plant or herb.

One impressive fact about plants is that they are naturally forgiving and easy to handle. With that in mind, one easy and effective way of gaining a full and deep understanding of the field of alchemy, as well as everything related to yourself and the galaxy, is to study and tincture plants. Ancient alchemists use seven heavenly bodies or planets along with the metals they are connected to when undertaking self-reflection.

Each planetary body, along with its corresponding planetary metal, strongly reflects the archetypal patterns created by the mind as a means of expressing consciousness. In essence, every planet in the realm of alchemy reflects one area or aspect of your mind. Such reflections are the ones that form your physical reflections, specifically those you are seeing as your physical body as well as your external reality.

In understanding alchemy, you can utilize the law of planetary correspondences, so you can find the relationship of something in your reality then comprehend the underlying theme or archetype of it. Note that our planet breathes life into many things, including plants, animals, and minerals. All metals on the planet that are considered as our foundation are also derived from stars.

Aside from being so dense, these metals also have their own unique elements. This means that they do not just consist of a set of complicated and incomprehensible molecules.

For instance, if you combine metal with another element, like sulfur, nitrogen, or oxygen, then it will create a mineral, an ore with metal being incorporated into its unique structure. There are also instances when minerals have multiple planetary correspondences, especially if several elements were incorporated into them.

Since the life of plants greatly depends on minerals, and minerals also exist while completely relying on metals, it is safe to say that plants have their own planetary correspondence. Every time you need to work with a plant or herb, you will most likely realize and comprehend the specific areas of the body where it will help medicinally.

You will also get to know its exact effects on your emotions and mind. The only thing you should do is to understand the planetary correspondence of such plants or herbs. Considering that, you can safely assume that you can find planetary metals in plants and extract or harvest them for your benefit.

A lot of herbal medicines are even recognized for doing a lot of good things for the body. The main reason behind these positive effects is the presence of multi-layered influences derived from the planetary metals. To determine the unique traits of various medicines developed through plants and herbs, you have to comprehend the planets' archetypes. Note that you can only grasp the specific planetary energies that work with a specific plant or material if you carefully and thoroughly study its inherent qualities.

Remember that trying to get to know more about the plant or herb's material is the key to identifying the planetary influences it has. You have to know its preferred conditions and environments, modern and traditional uses as far as health and medicine are concerned, its chemistry, and the mythologies linked to it.

Getting to know all these things will also help you understand the specific planetary metals associated with each plant and how you can make the most out of them by harvesting them and putting them to good use. You also have to know that the planetary association and correspondence of each herb usually comes from a couple of theoretical underpinnings—the traits of the plant and its medicinal action.

You can see the plant's specific nature being reflected by its traits, including the form through which it grows, the shape of the leaf, reproductive structures, taste, scent, habitat, and the color of the plant's flowers. You can also couple it with its actual effects on the human body. All these work in determining the specific planetary affiliation you can appropriately assign it to.

For instance, rosemary and other evergreen herbs with life force expression viewed as constant and steady—similar to the sun, can be categorized as solar. Plants with yellow flowers, like calendula and those that resemble the sun, such as the eyebright, also fall into the same category.

In that case, you can also expect them to deliver naturally solar medicinal actions—among which are strengthening your heart, clearing your eyesight, and making you feel warm, at ease, and generally good.

Let us use St. John's Wort as an example in that scenario. If it blooms during the midsummer season and comes with flower buds with amber-red fluid and flowers that are bright yellow in color, then you can consider it as an herb belonging to the sun. The same is true if you use this herb in promoting your wellbeing and dispelling anxiety.

You can also access and identify the planetary nature of a certain plant based on the specific place where it grows. For instance, herbs and plants that grow in graveyards, ditches, and other waste places are often viewed as the plants related to Saturn.

Overall, herbal or plant alchemy requires the thorough scrutiny and examination of a specific herb or plant to find and gather its planetary metals. With that said, it may take a long time to complete the entire process just to make the essences from plants and herbs achieve perfection.

To give you an idea, here are a few examples of plants and herbs used in plant alchemy categorized by the planetary metals that they provide.

Plants/Herbs with Planetary Metal Gold

Plants/Herbs (some examples)	Colors	Body Parts	Diseases	Drugs/Purpose
Acacia tree resin & seeds Arctic poppy Ash Blue water lily Buttercup bush Calendula Cedarwood oil, seed, and herb Chamomile oil, seed, and herb Citrus oil and herb Cinnamon Everlasting Hibiscus Japanese calamus Marigold Mistletoe Poinciana Saffron St. John's wort Sunflower Walnut Wartweed	Amber Golden yellow Orange Red Rich amber	Heart Eyes, specifically the right one Circulatory system Upper back Spinal column Blood Thoracic spine Spleen	Allergy Heart illness Sunstroke Dermatitis Fever Chills Skin rashes Eye disease Skin cancer Photophobia Hot flashes Substance abuse brought on by issues with self-esteem	Alcohol Antidepressant Tonic Warming drugs, though not as intense as planet Mars Red tincture Wealth acquisition Prosperity Protection

Associated with the sun, the planetary metal gold can be found in several plants and herbs. One thing that makes these sun (gold) plants good to use in herbal or plant alchemy is that they tend to promote general protection and prosperity. The sun also governs them, which is perhaps the most famous out of the planetary influences that are good for the general population. With that, you can expect the plants here to have the following qualities:

- Have resemblance to the sun in terms of color and shape.
- Tend to open to the sun then close at night.
- Promote warmth & relaxation.
- Serve as medicinal plants designed to have a positive effect on the heart.

You may want to use these sun plants containing the planetary metal gold if what you want to achieve includes centering and executing finance or money magic. You can also use such plants and herbs to honor all sun-related divine aspects. Most of the plants in this category are flowers shaped like the sun. One example is the daisy; you can also ingest them if you want to achieve calm warmth.

Plants/Herbs with Planetary Metal Silver

Plants/Herbs (some examples)	Colors	Body Parts	Diseases/Conditions	Drugs/Purpose
Agave Almond Banana Blue Hibiscus Blue Water Lily Cabbage California poppy seed Chamomile oil and seed Chickweed Clary sage oil and seed Cucumber Evening Primrose Flowering rush Green calla Heartleaf Hydrangea Iris Juniper essential oil and herb Lettuce Moonflower vine Mushroom Passionflower	Silver Blue Cold pale blue Sky blue	Lymphatic system Brain Stomach	Structural brain issues, such as tumor Seasonal Affective Disorder (SAD) Premenstrual Syndrome (PMS)	Hormone-balancing Sedative Tonic specifically designed for the stomach or brain Painkiller Narcotic White tincture Divination using dreams Clairvoyance Targets emotions

Rose Mallow			
St. John's flower			
Star anise essential oil and herb			
Tomato fruit			
Wallflower			
Water hyacinth			
Wintergreen			

You can also find plants and herbs containing the planetary metal silver, linked to the alchemical planet the Moon, such as the examples in the table. Plants in this category are received well by alchemists as most of them are ideal for use in divination, particularly when applied in dreams.

The majority of them have high water content, making them juicy and watery. Some examples are melons and cucumbers. This is the specific nature that makes them so moisturizing and cooling when used. Being related to the moon, you can see these plants have some parts shaped like a moon. They also have a smooth and sweet scent. Some of these plants are sedating while others are euphoric.

If you want to take advantage of these plants through herbal alchemy, then use them whenever you need to work with your emotions or your subconscious mind. These herbs and plants are also effective every time you intend to explore the astral plane. Moreover, you will find them useful as medicinal plants that can positively influence your sleep and emotions. They can also help improve the health of the female reproductive system.

Plants/Herbs with Planetary Metal Iron

Plants/Herbs (some examples)	Colors	Body Parts	Diseases/Conditions	Drugs/Purpose
Acacia resin Aloe American persimmon stem Artichoke thistle Ashwagandha Asian wild ginger root Basil Barberry root Belladonna herb and seed Black gum leaf Butterbur Cardamom Cassava leaf Chickweed* Chili pepper Chives Coffee senna seed Corn salad Cypress essential oil and seeds Dandelion leaf Garlic	Bright red Scarlet Azure Red Venetian Red Emerald	Muscular system Gallbladder Bile Blood Tendon Adrenal glands Genitals Face Head Arteries	Low energy and physical strength Poor immune system Accidents and injuries Low sex drive Immune or blood disorder Infection Rash Fever Stress and aggression Hypertension Ulcer	Absinthe Stimulate and heat tensed and forceful muscles Improves energy and strength Improves sexual drive and natural vitality Works against the influences of Venus and the Moon

Ginger
Mustard
Nettle
Onion
Pigweed leaf
Radish
Red clover shoot
Safflower blossom
Thyme
Turmeric
Wild ginger
Wormwood herb, oil, and seeds				

The plants mentioned here that contain planetary metal iron usually have prickles and thorns. The presence of these parts is unavoidable considering the fact that iron is represented by Mars, which refers to a soldier. The fact that Mars is an extremely physical planet is also the reason why the plants here have a spicy and strong taste.

As medicinal plants and herbs, expect them to work effectively in purifying the blood and boosting sexual potency. They also help to cure any problems with your head. In most cases, Mars (iron) plants also gained recognition for having heating and quickening properties.

You should consider using these plants or botanicals if you intend to create an elixir that is valuable for attack or defense. These plants are ideal for use on issues concerning masculine energy as well as when it comes to boosting courage and strength.

Plants/Herbs with Planetary Metal Quicksilver (Mercury)

Plants/Herbs (some examples)	Colors	Body Parts	Diseases/Conditions	Drugs/Purpose
Bladderwrack Butterbur California poppy seeds Caraway essential oil Cassia herb Celery Chickweed Clary Sage essential oil and seeds Dragonhead Fenugreek Garlic mustard Gobo root Irish moss Juniper berries essential oil and herb Lavender essential oil, seeds, and herb Lemongrass essential oil Mercury	Indigo Yellow Purple Gray Violet	Nervous system Brain Respiratory system Consciousness	Poor communication Inability to understand basic mental processes Irrational mind	CNS stimulant Herbs to improve the nervous system, brain, throat, and lungs Stimulates beneficial electrical activities within the body Nervine for balance

Mullein Parsley Russian sage Sweet Pea Woody nightshade			

Being ruled over by the planet Mercury, these plants or herbs that contain quicksilver are those that have a positive effect on the nervous system as well as one's thoughts. You can often see these plants or botanicals with finely divided and feathery leaves. Most of them also deliver a zingy and high scent, though it does not seem to last that long. The reason is that Mercury has quickening properties.

You can find medicinal plants in this category known for being good for your brain function, thought, nervous system, and speech. The herbs here also have properties designed to improve memory and intellectual capacity.

The good thing about these plants and herbs is that they also seem to work well with others meaning you can also use them with other botanicals to further improve their potency. You may want to use these plants every time you intend to learn and practice new rituals, show your glamor, or communicate effectively anytime you find yourself in a difficult situation.

Plants/Herbs with Planetary Metal Tin (Jupiter)

Plants/Herbs (some examples)	Colors	Body Parts	Diseases/Conditions	Drugs/Purpose
Anise Barley stem Bayberry bark Bell pepper fruit Bitter aloe leaf Blessed thistle Brazil nut seed Carrot root Catnip Chicory Chinese ginseng root Clove Pink Corn kernel Dandelion root Echinacea root Fennel seed Fenugreek seed Gobo root Hydrangea Juniper berries Licorice root Marshmallow root Nutmeg	Bright blue Violet Yellow Blue Rich purple	Digestive system Liver Metabolism Body's building and anabolic processes Arteries	Poor nourishment Ailments caused by poor judgment, such as anorexia and alcohol addiction Fatty tissues	Cocaine Laxative Tonic designed to stimulate your appetite Herbs with effects on weight Promotes optimism Improves approach to life and when making judgments Useful in faith, law, and religion

Peppermint leaf			
Sage leaf			
Stevia leaf			
Walnut			
White willow bark			
Wild yam root			

Other sets of plants and herbs used in herbal alchemy are those with the planetary metal tin, which is also linked to Jupiter. Tin (Jupiter) plants are often bold and large. In most cases, this group consists of seeds and fruits, like figs, olives, and nuts, containing many nutrients. You will also notice some of them having yellow flowers because yellow is one of Jupiter's colors.

Another distinctive feature of the plants in this group is that they are aromatic, though not similar to the aroma produced by plants in Mercury or with quicksilver. When used, the plants can provide warmth. It is as if you feel a sense of sedate contentment without stupefaction or sleepiness.

You can also use these plants anytime you wish to feel relaxed, warm, calm, and expansive. If you want to hold a ritual with a main focus on law, faith, authority, and religion, then you may also find the plants and herbs with this planetary correspondence and influence useful.

Plants/Herbs with Planetary Metal Copper (Venus)

Plants/Herbs (some examples)	Colors	Body Parts	Diseases/Conditions	Drugs/Purpose
Almond Angelica Apricot Asparagus pea seed Bergamot Mint Bitter melon fruit Black cherry stem and leaf Blackberry Broccoli leaf Cabbage leaf Cashew seed Clover Collard leaf Cucumber fruit Daffodil Dwarf sumac stem Eucalyptus Garden sorrel leaf Geranium essential oil Honeysuckle	Bright rose Emerald green Early spring green Sky blue Pale green Cerise	Kidneys Genitals Urinary tract/system Hormones Puberty Sex Bladder Prostate Testes Ovaries and uterus Facial skin Tongue Affection, sources of pleasure	Kidney and urinary problems Reproductive and sex problems Skin issues Emotional upset Grief Depression	Aphrodisiac Tonic and cleansing herbs for the skin Tonic for urinary tract and kidneys Provide balance to the female system

Jerusalem artichoke				
Lettuce leaf				
Moneywort				
Natal cherry				
Peach fruit				
Pignut hickory shoot				
Plum fruit				
Pumpkin leaf				
Red mangrove leaf				
Rosehips				
Sesame				
Sycamore				
Tomato fruit				
Willow oak stem				

Another group of plants and herbs perfect for alchemy are the ones that contain the copper metal. With its corresponding planet, Venus, copper-rich plants and herbs used in the field of alchemy often include lush and huge flowers that have powerful scents. Some plants in this group also have soft and furry leaves, red fruits, and beautiful flowers with a sweet scent. Most of these plants are also mildly astringent.

You may want to use these plants anytime you intend to conduct a ritual linked to love and money. You can also use the planetary influence delivered by this plant to combat martial spells. Moreover, these plants and herbs work well for those who have a strong desire for harmony and those who intend to acquire valuable things and improve their talents and skills.

Plants/Herbs with Planetary Metal Lead (Saturn)

Plants/Herbs (some examples)	Colors	Body Parts	Diseases/Conditions	Drugs/Purpose
Aconite root and seeds Asparagus shoot Belladonna herb and seed Blackbean fruit Black gum stem and leaf Black oak stem Buckbush stem Comfrey Corn seed Cypress herb, oil, and seed European nettle leaf Globe amaranth Ironwood Lettuce leaf Black nightshade Northern red oak stem Pignut hickory	Indigo Black Blue Amber Dark brown	Excretory system Bones Teeth Muscles	Phobias and Obsessive-Compulsive Disorder (OCD) Ulcer, hypertension, and other ailments triggered by the strong urge and need for control Problems with discipline, administrative ability, and organizational skills	Cooling, stabilizing, drying, and constricting herbs Herbs and tonics for the spleen as well as body structures and their weaknesses, like bones and arthritis Solution for aging Deliriants or hypnotic herbs

shoot			
Post oak stem			
Rupturewort			
Shagbark hickory shoot			
Smooth sumac stem			
St. John's wort			
Tomato fruit			
Wall fern			
White oak stem			
Willow oak stem			
Witch Hazel			

The plants and herbs that have the planetary metal lead are usually all about understanding limits and highlighting borders and limitations. Governed by Saturn, expect these plants to grow in dry grounds incapable of nourishing other plants. Some of them grow in the shade, too. Sorcerers and alchemists from the past to the present use these plants when making hexing or binding spells and rituals designed to break unwanted habits.

Apart from being constricting, stabilizing, drying, and cooling, the herbs here also aim to rejuvenate and slow down or fight the aging process. You will also notice some of these plants having pale and wispy flowers and producing plenty of seeds. The fact that the planet Saturn is kind of slow also means that the plants in this group grow slowly, too.

Chapter 8: The Spagyric Process

Spagyric is synonymous with alchemy. It follows the same procedure used by alchemy, which involves the purification of objects and making them attain perfection. Eventually, though, the spagyric term no longer covers alchemy in a general sense. It has become more specific as it started referring to the herbal medicines that the alchemical procedures produce.

In other words, it became more connected to the operations and processes used in herbal alchemy. Among the operations involved in the spagyric process are fermentation, extraction, and distillation of mineral components taken from plant's ashes. Such procedures were used during medieval alchemy as a means of separating and purifying metals from ores, as well as salts from aqueous solutions, including brines.

Spagyric as Part of Herbal Alchemy

The spagyric process often results in the development or creation of herbal solutions or remedies. For this reason, it is now more connected to the field of herbal alchemy instead of alchemy in general and its entirety. In that sense, you can consider the spagyric process as both an art and science. It ferments, extracts, purifies, and recombines plant matter with the ultimate goal of producing medicinally potent solutions and substances.

Spagyric alchemy also makes use of the three basic properties of alchemy in a specific plant matter. These are mercury (water), salt (earth), and sulfur (fire). The entire process involves dividing the raw material first into a few components. After that, it separates the useful or valuable components from the ones you can't use. You will then be required to use a kind of re-composition, which is helpful in combining visible parts.

As you can see in the basic steps mentioned, one goal of spagyric is to separate the energy-giving, healing, constructive, and good principles and components of each substance or plant that are ideal for use in healing, from the destructive, pathogenic, and unhealthy toxins. Spagyric also tries to enhance such good principles through supplementation.

With the meticulous way of forming spagyric substances, including metal essences and herbal drops, it is safe to say that such results have the most potent and powerful properties. The good thing about all of them is that they are in their enhanced form, so expect them to work even better for therapeutic and healing purposes than raw materials.

The human body can also easily digest them, because they are already devoid of toxins. Apart from having a positive influence and impact on your physical well-being, the substances produced by the spagyric process also work to boost your mental and spiritual health.

Make Spagyric Tinctures and Substances

You can prepare spagyric tinctures and substances using various methods—among which are different fermentation and distillation procedures. A basic process that you can carry out is that involving the production of a normal tincture using subsequent mineral salt recombination.

To collect a specific plant's mineral salt, the plant has to be macerated in alcohol first. You should then calcine the remaining plant material until you notice it forming into white ash. When that happens, expect the salt to be leached out from the white ashes and purified.

The next thing to do is to crush the purified salt into a fine powder. The goal is to recombine it with the tincture. You are also allowed to recombine it with the mercury and sulfur you have previously extracted. This can leave you with a product, which uses the entire plant.

So basically, you can divide the entire spagyric process into the following steps to make it easier to understand:

- Alcohol extraction, which extracts plants with the aid of organic grape alcohol that you can place inside a Soxhlet extractor. It can help produce high-potency tinctures quickly. One thing to note about this procedure is you have to do it under a vacuum. With that in mind, you have an assurance that any heat-sensitive part of the plant will not be destroyed.

As an alternative to the method, which involves the use of a Soxhlet extractor, you can do the extraction using a 40-day maceration and a jar. This method requires you to dissolve the plant's essential oils and alcohol soluble chemicals. This approach can help you to extract not only the oil-soluble compounds but also water.

- Mineral extraction, which involves pouring the tincture then saving and storing it in a dark spot. It is the step where you have to burn the remaining plant material you have extracted then calcine it to extract all the minerals present in the ash.

After you have formed white ash out of burning the herb or plant, it is now easy to dissolve the mineral salt readily in water. You can then combine the ash with distilled water as a means of dissolving all the minerals. Afterwards, filter the water, allowing it to evaporate slowly, which is helpful to crystallize mineral salts.

- Recombination, which involves gathering and recombining the white salts with the previously created plant tincture. This step will complete the entire spagyric procedure.

After the last step, the tincture can be expected to have all the oils, minerals, and acids which were originally present in the plant. The only difference is that it was processed to produce a more purified and potent variation of it.

With that said, you will notice the plant's chemical aspects and energetics being taken to the maximum level. The completion of the spagyric process also means that the new compounds formed are specific to that plant's chemistry and its exact preparation method.

Find Plants and Other Supplies for the Spagyric Alchemical Operation

Your decision to work with spagyrics can give you the chance to get yourself attached and connected deeply with your chosen plants. Once that happens, expect their positive subtle, and strong effects not only on your body but also on your mind and spirit. The good thing about this specific alchemical approach is that it combines the entire plant's elements both energetically and medicinally.

You will also notice the medicinal compounds becoming more bioavailable, which means that your body can readily absorb them. It also helps in increasing the activity of your body. What is even better is that it is already possible for you to get a complete medical dose with just several drops, often around five to ten drops.

The pagyric process also helps in creating an elixir or tincture, which, when taken, will let you receive the plant's deep wisdom in your soul, body, and mind. Another advantage of this process is that it assures you that you get to extract the raw material you are using at the maximum level.

Aside from that, it gives you a guarantee that you will be able to take full advantage of all the essential healing parts of the plant minerals often found in the pulp. You can get even those that are only available through extraction. One more thing that it can do is create a more concentrated product compared to the original raw material.

Note, however, that you can only take advantage of such positive effects if you source the right plants and supplies for the entire spagyric alchemical operation. Some tips that will let you access the right plants and supplies and ensure that you will be doing the spagyric process on a budget are as follows:

Look for Fresh Plant Materials Instead of Dried

Generally, it would be a lot better for you to pick fresh materials for the specific plant you intend to use over the dried ones. If you need to use a dried herb, then it means that you should buy a fresh one and do the drying on your own.

It can assure you that what you are getting is genuinely fresh, not a dried variety that has already been oxidized by staying on the shelf for several years. The latter is also low in terms of quality, so be extra careful. Note that low-quality plant materials may cause your output to be not on par with the quality you initially expected.

With that in mind, always shop for fresh and high-quality plant materials. It is a big help in ensuring that you will get your money's worth. Another tip is to acquire a few identification books designed for plants grown in the bioregion of your locality.

Aside from giving you the chance to locally harvest the plants, you get the opportunity to familiarize yourself with the local habitat and the medicines growing in your own environment. If possible, look for an instructor in your locality who can assist you in identifying local medicines appropriately.

Identify the Herbs You Intend to Work With

It is also a must to identify all the herbs you are working with or still planning to work with. The goal here is to prevent yourself from making a dangerous mistake when you use an inappropriate herb or plant. Be clear with the identification by using books relevant to your purpose. A wise tip is to look for and use books designed for your specific locality.

Spend time researching organic herb growers and farmers who supply fresh herbs and plants in your area, too. For plants and herbs that require shipping, get a guarantee that they will be appropriately and securely packed using ice packs. That way, you will never have to worry about losing their freshness.

You may also want to negotiate with and talk to smaller producers who you can coax to harvest plants and herbs based on planetary timing mechanisms. Expect most of them to feel excited doing that, plus you can lower the amount you may have to pay in the end.

Look for the Best Alcohol Provider

It would be wise to buy the alcohol offered by Alchemical Solutions, now called The Organic Alcohol Company. This company can supply you with pure organic alcohol and at a reasonable price. One advantage of this alcohol is that it comes from various sources such as cane, grapes, and grain.

For spagyric alchemy purposes, the best alcohol that you can use is probably that which is derived from a grape. One reason is that grape alcohol produces the highest quantity of distilled spirits or volatile mercury, making it hold the highest rank in terms of the plant kingdom evolution. The problem is that it is also quite expensive, but it is worth it, especially if you consider its overall quality.

Make Your Distilled Spirits

You can also do the spagyric alchemical procedure on a budget by deciding to create the distilled spirits that you can use for extraction on your own. You can do it from scratch. All you have to do is buy plenty of grapes then use them to produce your own wine. Once you have made the wine, you can distill it until it is 95 percent alcohol.

Doing this process on your own offers you a fantastic experience, as it allows you to experience what it feels like to do quality fermentation and distill alcohol. You may also want to shorten this DIY procedure by buying a wine of top-notch quality. The only thing that you have to do in that case is to distill it.

Another advantage of this DIY approach for extraction is that it lets you distill alcohol based on the actual planetary timing mechanisms. With that, you can infuse it based on the planetary influence you intend the medicine to have. It means that what you will be preparing perfectly fits your standards while still ensuring that it does not deviate from the spagyric alchemy principles.

Use Water

Instead of using 95% alcohol when doing the spagyric procedure, you may also want to dilute it first with water. It is beneficial because it can produce the highest extraction quality of the specific plant you are working with. You can also use water during the mineral salt dissolution process. It is just a simple process that uses the plant's hydrosol. Alternatively, you can use the plant's phlegm in the absence of the hydrosol.

In case you plan to make a spagyric tincture that does not require any distillation procedure, a great recommendation is pure distilled water. You just have to mix it with the 95 percent alcohol so you can go through the extraction process. You can choose to buy the distilled water you will be using for this procedure.

However, if you want to save while learning simultaneously, note that you can distill the water on your own. Just make sure that you do so based on the planetary timing mechanisms suitable for the tincture or remedy you wish to create.

It is the best way to incorporate planetary energies into the menstruum and the extraction timing itself. This can result in improving the medicine's subtle astral traits. If you decide to use water, then you will surely be glad to know that there are various methods to collect it. You can collect spring water directly from the earth and some sources of it in your locality.

You can also use rainwater. It is said to be perfect for spagyric procedures. You can produce even better results if you get such rainwater from a thunderstorm. Also, remember that dew has a major contribution to the field of alchemy. With that, you can also collect it during springtime for use in your spagyric procedures.

Gather All the Other Supplies You Need

It is also crucial to be aware of the specific supplies you will be needing for the spagyric procedure before starting. That way, you will have an idea about what you should invest in. In this case, you may want to check out the following supplies and equipment that a lot of alchemists use during the course of making their own spagyric solutions.

- **Lab equipment/glassware.** One supply in this category that you should invest in if you are serious about doing spagyric alchemy is the basic still, an apparatus to distill liquid. It is a big help for sulfur extraction. You can get one for around $300 to $400 or less. Just make sure that you check the seller's reputation so you can get a truly high-quality product that will let you distill oils immediately.

It is also advisable to invest in set-ups that are all friendly to the spagyric alchemy process. Among the supplies you will need for such a set-up are large Soxhlet extractors, circulation and concentration vessels, vacuum distillation set-ups, large stainless-steel distillers, and pelicans.

One of the best sources of all the mentioned laboratory equipment and the glassware is a legitimate online marketplace, like eBay. The good thing about buying this stuff online is that it will cost you less. Just make sure that you spend time researching the items offered to you online to find out if those are still new and ideal for use in processing harsh chemicals.

You can also buy heating sources from eBay at reasonable prices. There are even those that cost not more than $500. You just have to make sure that the products you are getting are all high-quality and can help give you the kind of output you want.

- **Essential supplies for fermentation.** One of the things you may need is a fermentation vessel, which you can buy at a brewing or fermentation supply shop in your local area. It would be ideal to use a glass-based one. Another option is a bucket specifically made for the fermentation process. It should be certified food-safe. Avoid plastic-based vessels as such material is not particularly good for the purpose.

You may also need airlocks. Aside from the actual fermentation vessel, you also need to prepare other items that you can use for this step, including rubber stoppers that come with drilled holes. With these holes around, you can easily fit the airlocks.

Other things that you can use for fermentation are high-quality wine yeast and sugar. When using sugar, make it a point to choose fructose because of its endogenous nature. You can also buy fructose from Steviva, which is a great provider of the product.

One advantage of this company is that it provides its crystalline fructose, known to be non-GMO, at low prices. If possible, purchase this product in bulk so you can make the highest possible savings.

- **Tincture press.** Another supply you should invest in is a tincture press. Many consider it as the most important laboratory supply or equipment. It is a hydraulic press, which works in squeezing a lot of liquid from plant materials. The good thing about using this tincture press is that it can maximize your use of the medicine you produced.

You can check out different sized tincture presses, so you can choose one that fits your scale. Among the best sources of tincture presses are Strictly Medicinals and Longevity Herb Company.

Besides gathering all the supplies and ensuring that what you are getting is of top-notch quality and reasonably priced, it is also advisable to arm yourself with all the information about the spagyric alchemy process. That way, you can continuously work on making tinctures, remedies, and elixirs.

In that case, you may want to start reading books with all the sources of information about spagyric. This is to prevent you from jumping in blind just to take advantage of spagyric herbal alchemy. Some of the books you may want to check out, in this case, are *The Way of the Crucible* and *Real Alchemy*—both of which are from the famous author Robert Allen Bartlett; *The Alchemist's Handbook* by Frater Albertus; and *Sorcerer's Stone: A Beginner's Guide to Alchemy* and *Emerald Tablet: Alchemy for Personal Transformation*, which were both written by Dennis William Hauck.

Chapter 9: An Alchemical Cookbook

Plant or herbal alchemy, which can now also be associated with spagyrics, serves as your entry point to the world of alchemy. The good news is that this specific form of alchemy allows you to work with plants and herbs that are all safer and easier to deal with compared to toxic metals, like antimony.

Another thing that you will enjoy from herbal alchemy is that there are several recipes that you can make. You can also easily access numerous herbs and plants, which you can use to create potions, flasks, concoctions, elixirs, and oils or while performing transmutations.

In this chapter, you will learn a few simple recipes that you can use to familiarize yourself with the entire herbal alchemy process. These herbal alchemy recipes will jumpstart your journey toward becoming an expert alchemist.

Herbal Infusions

Herbal infusions are among the things you can create out of plant or herbal alchemy. Here, you have the option of using various parts of the plant—among which are the leaves, flowers, and stamens, otherwise called pollen stems. The herbal infusion process somehow resembles the tea preparation procedure.

Note that the mentioned parts in plants or herbs are quite delicate or sensitive. It is highly likely that they will get destroyed if you boil them directly in water. You can preserve their benefits by putting them in a jar or mug first, then pouring the boiling water over them. The strength of the output usually depends on how long you infuse the parts of the plant.

The good thing about herbal infusions you can make is that they are so versatile that you can use them in various ways—among which are drinking them for their medicinal benefits, integrating them in homemade remedies and cosmetics, and making them part of organic and natural home products, like insect repellents and fertilizers.

Prepare Herbal Infusions

One advantage of herbal infusions is that they need less preparation time. The only thing you have to wait for is the thorough steeping of the herbs. In most cases, you will have to follow this basic recipe:

Ingredients
- 1 tbsp. of your preferred dried herb
- 1 cup boiling water

You will also need a clean glass jar that has a tight lid for this herbal infusion recipe.

Instructions

1. Put the dried herb you intend to use in the glass jar or container.

2. Pour the boiling water over the dried herbs. Make sure that all the herbs are fully covered.

3. Cover and seal the container using its tight-fitting lid. This is crucial in preventing the volatile oils and steam from coming out.

4. Let the infusion steep for a while. It should be until the added water cools down to room temperature. You can also increase the infusion time based on what is recommended in your chosen recipe. If you use the bark and roots of the herbs, you may need to infuse them for up to eight hours.

If you are using leaves, then the infusion should take at least four hours. For seeds, you will need a minimum of thirty minutes to infuse them. Meanwhile, the infusion time required for the flowers of certain herbs is two hours.

5. Use a fine-mesh strainer or cheesecloth to strain the herbs from the water. Do this step continuously until you have taken out all the herbs. This should result in a liquid, which you can now refer to as the actual herbal infusion.

6. Clean the glass jar and container, then pour the resulting infusion into it. Store the prepared infusion. You can refrigerate it for a max of 48 hours. Afterward, you will have to discard it.

Additional Notes and Tips

You can proportionately increase the amount of water and herbs that you will be using for the recipe. It is highly advisable to increase the quantity if you want to produce a larger amount of infusion. For instance, if you intend to prepare one quart of infusion at a time, then you may need around one cup of the herb and 1 quart of boiling water.

If you are still a beginner, it would be best to do small volumes at first. It ensures that you will not be wasting any of the infusion, just in case you do not like the result. Also, do not forget to cover the container or jar all the time so the steam is contained. It is necessary to keep the heat trapped inside, as it may help to release the beneficial compounds present in the herbs you are using.

Note that you can also make a decoction, which is a liquid or beverage stronger than the infusion. All you have to do is use the bark, roots, dried berries, and the other parts of the plant needing stronger and longer heat for oil extractions.

Here's how you can make the decoction if you prefer to make one:

- Mix your chosen herbs as well as the boiling water in a saucepan.
- Use the lid of the pan to cover it. Simmer the mixture slowly and gently. This should take around 20 to 45 minutes.
- Take the decoction out of the heat. Strain, then pour the resulting liquid into a jar or container for storage.

If you are unsure what herbs to use in your infusions or decoctions, then note that to begin with, the best choices include aloe vera, thyme, mint, sage, chamomile, red clover, nettle, lavender, and echinacea. They are great to use in your recipe because of their medicinal benefits.

Herbal Tinctures

Herbal alchemy also makes it possible for you to create herbal tinctures. What is great about these tinctures is that you can prepare them using all parts of the plants. The preparation of a herbal tincture involves the extraction of the vital properties of the plants or herbs using the following items:

- **Alcohol.** One of the ingredients you can use for this recipe, which can help extract the medicinal properties of the plant, is alcohol. It is a powerful solvent, making it effective for that purpose. It can extract the alkaloids of your chosen plant or herb. Alkaloids refer to the organic compounds housing several medicinal benefits.

Another advantage of using alcohol is that it is capable of preserving the prepared tincture for an indefinite period of time. It also mixes perfectly with water and glycerin. However, the problem with alcohol is that it is not that effective for extracting resin. It may even destroy specific compounds and may cause problems for those who are alcohol intolerant.

- **Vegetable glycerin.** Vegetable glycerin is also another possible ingredient you can use for your herbal tincture. It is an incredible ingredient to use because aside from being non-alcoholic, it also seems to be effective in resin extraction. The problem is that the tincture it can create is a bit weaker compared to the ones using alcohol. The reason is that it is not as strong as alcohol.

- **Apple cider vinegar.** You may also want to use apple cider vinegar for your herbal tincture. The reason is that it further enhances the benefits that the output can offer. Having apple cider vinegar in the tincture may contribute to neutralizing the pH level of your blood, supplying your body with antibacterial benefits, and reducing your cholesterol and blood sugar levels.

The problem with apple cider vinegar is that its strength is also lower compared to alcohol, especially when it comes to the extraction of medicinal compounds. Moreover, its shelf life is not indefinite, making it a bit inferior to alcohol.

Instructions

1. Choose from alcohol, vegetable glycerin, or apple cider vinegar. This should determine what you will be using to extract the properties of the plants or herbs.

2. Prepare a glass-based mason jar and put the plants you intend to use in there. Pour several tablespoons of boiling water. The water should be enough to open up the plants and prepare them completely for the process of extraction. As a guide, you can follow the plant to liquid ratio of one to five if you are using dried herbs. If what you plan to use in this tincture is a fresh herb, then the most suitable plant to liquid ratio should be one to two.

3. Fill up the remaining mason jar or container using your choice from alcohol, organic vegetable glycerin, and apple cider vinegar.

4. Seal the mason jar. Afterward, leave the herbal tincture for a minimum of two to three weeks so you can infuse it properly. Do not forget to agitate this tincture every now and then and store it in a warm place.

5. Strain the plants. Do this step if you are satisfied with the strength of the tincture after infusing. You can then put it back in the jar. Do not expose it to direct sunlight and use it whenever needed.

If you want to create a stronger herbal tincture, then make it a point to lengthen its infusion time. Another great example is a tincture that uses rosemary. You can prepare this rosemary herbal tincture by using the following:

Ingredients

- A handful of the branch of rosemary—it is also equivalent to one-half cup per one cup of liquid.
- Your choice between high-proof alcohol, like brandy or vodka, and organic apple cider vinegar.
- Organic and fresh rosemary.

Instructions

1. Take the branches from the rosemary. Chop these branches, then put them in the sterilized jar/s you prepared.

2. Pour the water or liquid over the rosemary branches in the jar. Seal it for around six weeks to a year.

3. Strain the leaves of rosemary. Do this step only if you have already infused the herb-base of your preference. In most cases, the minimum infusion time for this recipe is at least six weeks.

4. Store the jar in a dark and cool place.

Once done, you can finally take hold of a rosemary herbal tincture that you can use whenever you feel like you need it. You can take several drops of this tincture beneath your tongue whenever you feel sick and require a boost in immunity.

Herb-Infused Oil

You can also prepare a herb-infused oil from the plants and herbs that you use in herbal or plant alchemy. You can use the herb-infused oil for a couple of purposes—one of which is for cooking so you can bring out a new flavor for a particular dish. You can use it in a similar way as cooking oil, like regular olive oil. You can even use garlic oil, which is an herb-infused oil you can add to kitchen recipes.

You can also take advantage of herb-infused oils by adding them to your personal care and medicinal products. It is even possible for you to infuse the herb in your favorite tea in oil so you can enjoy its topical benefits.

When making an herb-infused oil, remember that you will need a couple of ingredients:

- **Dried herbs.** It is crucial to use dried herbs as those with moisture may result in a rancid and moldy output. You are still allowed to use wilted or fresh herbs, but you will have to go through the process of heating them first using low heat. Heat causes the herb's moisture to evaporate.

- **Carrier oil.** This is where you will be infusing the herbs you intend to use. Some carrier oils you can use for a herb-infused oil are olive oil, sweet almond oil, avocado oil, and coconut oil. Just make sure to choose the organic variety to ensure that the output will be of top-notch quality.

One basic recipe that you can follow for a herb-infused oil is this:

Ingredients

- 1 cup of your preferred herbs
- 1 and ½ cups carrier oil

Instructions

1. Put the dried herbs in a jar.

2. Pour the chosen carrier oil over the herbs. The oil should cover the herbs and be sufficient to fill up to around an inch of the top part of the glass jar.

3. Put the jar in a window exposed to the sun. It should be there for around four to six weeks.

4. Strain the herbs from the oil, then store the output in a dark and cool spot.

If you want to speed up the infusion process, you can heat the ingredients in a double boiler or slow cooker. You can use low heat and wait for three to four hours for the process to be completed. Strain the herbs, then store the oil in a cool and dark place.

Other Herbal Alchemy Ideas That You Can Prepare at Home

If you want to delve deeper into the world of herbal alchemy, then trying to make these recipes in the comfort of your own home can also help you to master the craft:

Basic Alchemy Oil

Ingredients

- 1 tbsp. each of turmeric, paprika, and coriander (all should be in powdered form).
- 1 tsp. powdered cinnamon
- 2 tsps. powdered mustard
- 1 and ½ cup olive oil
- ¼ tsp. salt
- Optional ingredient: 1 tsp. vodka. Ensure that you go for the 100-proof variety.

Instructions

1. Process the dried herbs in your blender so you can form a powder out of them. You may also want to use the pre-powdered versions of them. You just have to purchase them from a reliable provider.

2. If you plan to use the vodka, you should do the alcohol rehydration process next. What you should do is place the powdered herbs inside a bowl. Rehydrate them using vodka by mashing and stirring the alcohol into the powdered herbs.

Do it until there are no clumps in the bowl. Let it sit for half an hour while covered. This step, despite being optional, is a big help to improve the process of extracting the savory flavors and medicinal properties of the herbs.

3. Get a sterilized jar and place the herbs together with olive oil and salt there. Shake well.

4. Allow the mixture to sit for around two weeks. After that, you can use the oil.

What is great about this recipe is that it no longer requires straining. You can just let the spices stay in the oil. Just make sure that you shake the container before each use, so the herbs will be able to mix properly into the oil.

Another advantage of this infused oil is that you can use it for up to a year. You can preserve its shelf life by refrigerating it. Also, keep in mind that excess heat, moisture, and heat exposure, as well as fluctuating temperatures, may hamper the oil's integrity, so ensure that you put it in a place where those issues are not present.

Frankincense or Myrrh Tincture

Ingredients

- 30 grams or 1 ounce of fresh and finely powdered myrrh or frankincense oleo gum resin
- 150 ml. vodka

Instructions

1. Prepare a resealable and clean glass vessel, then put the fresh and finely powdered myrrh or frankincense oleo gum resin in there.

2. Pour your chosen vodka. Make sure that it is unflavored and is either 80 proof or contains a minimum of 40 percent alcohol.

3. Let a small amount of the oil run on your finger. Do it around the jar's thread, almost to its lip. This is a crucial step in ensuring that the resin seeping in through the capillary action every time you shake the tincture does not seal the lid closed.

4. Tighten the lid by hand afterward, then store it in a relatively warm area.

5. Shake it at least once every day to dislodge all the material from the glass vessel.

6. Continue doing the maceration for a minimum of four weeks or one full moon cycle.

7. Take the tincture out of the glass once you are one hundred percent sure that there is no longer any color that transfers to the liquid or menstruum. After that, pour the resulting tincture through a filter. It should be fine enough, like a paper coffee filter.

8. Fold over the paper's edges on to the material that is now most likely exhausted. Use the back part of a spoon to press it gently and remove any leftover moisture.

9. After that, cover the filtered tincture, then allow it to sit undisturbed for one to two days. This should be enough time for the tincture to settle and sediment.

10. The next step is pouring or siphoning off the clear liquid so you can transfer it into a clean and sealable jar or bottle. Add appropriate labels.

Make sure to store this tincture at the appropriate place, preferably in a spot where it does not get overly exposed to the direct heat of the sun, as well as other factors that may affect its quality, effectiveness, and shelf life.

Chapter 10: Becoming a Modern Alchemist

Now we are on to the last chapter of this book, which tackles some of the tips and rules you should follow if you want to turn yourself into an effective modern-day alchemist. By now, you have probably grasped what this field is all about. You may have already understood all alchemical processes and theories, thereby encouraging you to persevere with this goal.

However, just like any other goal in life, your decision to become a modern alchemist requires the right mindset. It is also crucial for you to stick to an alchemical lifestyle, which applies various elemental energies. That way, you can become a more effective alchemist.

It is also advisable to keep in mind the following tips and tricks for modern alchemists, so you can apply them once you begin your endeavor in this field.

Break It Down Yourself

Traditionally, calcination is all about breaking down a substance by exposing it to heat in a crucible. Psychologically, what you have to break down is your sense of self. The goal here is to change yourself into someone who is stronger and more aware. Think of your old self as the lead you have to transform. You should break it down so you can transmute yourself to reach your gold potential.

Spend a minimum of twenty minutes daily for reflections—you should also contemplate what is happening in your daily life. That way, you will know exactly what areas in your life you should change. Choose a self-reflection method that is most convenient for you and turn it into a habit—one that you should be able to do without qualms daily.

Perform Rituals Inside a Magical Circle

One thing that you can do is to visualize a magical ritual—one that turns a typical room into a doorway with magic through specific clothes, symbols, candles, and decorations. There should be a magical circle in the room, which is outlined on the floor, where you can perform the ritual.

It is crucial to perform your rituals with such a mindset as to get the result you want. Note that a lot of alchemists and magicians believe that doing rituals inside a magical circle, which is present in a well-prepared room, can help to transmute thought energy to the physical realm or world. They just have to make sure that it is also combined with the appropriate and proper intent.

Accept and Believe the Power of Alchemy

Remember that you will also have a high chance of becoming a modern and expert alchemist if you have strong faith in this field. Remember that the key to achieving success as an alchemist always includes acceptance and beliefs. What you should do is examine your own life. Remember all the rituals you have participated in. You should then observe all areas and aspects wherein real-life magicians can create alchemy successfully through intent and ritual.

Remember the things they did to produce great results for the people they decided to work for, as well as the thousands or millions of consumers who accepted and believed their magical creations, particularly those they willed them to exist.

You have to accept and believe other magicians through your thoughts. It is also necessary to accept and believe in your own skills. That way, you can confidently perform rituals that have a higher chance of producing the results you want based on your intentions.

Moreover, acceptance and belief should be practiced when looking at your subjects. This means that you should embrace the fact that your subjects accept and believe your alchemical creations.

Remember the Importance of Agreement

Another important thing that you should inculcate within yourself if you want to maximize your full potential as an alchemist is agreement. You should recognize that agreement is what you need to get the power you want in the world of alchemy. You can begin the agreement mindset with the thought of using your will to pull yourself from the ether. This means having an agreement with yourself that you can manifest whatever it is that you are thinking into reality.

It also involves learning the moves of the most famous magicians—the ones who have already proven their mettle in the craft of magic. You have to learn and practice from them, so you can refine your rituals. Doing so can also help you reach an agreement with others that your alchemical creations are real.

Remember that transforming yourself into a powerful alchemist requires you to be a master when it comes to agreement. This means that every reality that you would like to achieve will depend on other people agreeing that they will give it to you. It could be love, fame, friendship, money, influence, and even sex. In other words, you also have to be a negotiator if you want to succeed in modern-day alchemy.

Continue to Learn

Being a master in alchemy is also all about your drive to learn everything about this field. You have to motivate yourself to continue learning instead of just stopping at the traditional theories. Your goal should be to grow as an alchemist by learning every bit of information about alchemy, especially the new tips and tricks.

You also have to try educating yourself by learning from others. This means you should get a full understanding of other people's game so you can get a trick or two from them. Moreover, you have to study the methods used by other magicians. It does not matter if the method is good or bad. What is important is that you learn from the others in this industry.

You should then take this knowledge and use it in such a way that it will improve the value of your life and the lives of other people. When trying to learn alchemy, avoid manipulating materials—whether metals, plants, or minerals. Note that alchemy comes with its own unique matter, which is the primary secret key you have to comprehend. Instead of merely manipulating materials, it helps to delve deeper into the secret of alchemy so you can practice it even better.

One more thing you should practice is patience. Remember that you will most likely face ups and downs as an alchemist. You are even at risk of spending several years getting frustrated with the failures you will encounter when doing alchemical operations. Do not give up when that happens. Continue practicing, and eventually, you will get to understand completely what makes the science of alchemy so divine. When that happens, you can see yourself doing better and better as an alchemist.

Conclusion

Alchemy is indeed one of the most fascinating ancient practices still in existence today. You can learn from this craft and turn yourself into an effective alchemist. Just make sure that you commit to the specific goal that you have in mind when trying to study alchemy.

Learn from the masters—the ones who have already proven how good they are in performing alchemical operations. It is also advisable to recognize their game so you can figure out which of their methods you can apply on your own. Moreover, it is crucial to refining your rituals by constantly educating yourself about alchemy.

Hopefully, this book gave you all the important information about alchemy to help you understand this field better. Use all the things you learned here to improve your potential as an alchemist and make all aspects of your life flourish. Your goal should be to do everything correctly so you can use alchemy to add value to your life and help yourself and those around you to achieve genuine happiness.

Here's another book by Mari Silva that you might like

Your Free Gift (only available for a limited time)

Thanks for getting this book! If you want to learn more about various spirituality topics, then join Mari Silva's community and get a free guided meditation MP3 for awakening your third eye. This guided meditation mp3 is designed to open and strengthen ones third eye so you can experience a higher state of consciousness. Simply visit the link below the image to get started.

https://spiritualityspot.com/meditation

References

12 Alchemy Symbols Explained. (2018, August 20). Ancient Pages. https://www.ancientpages.com/2018/08/20/12-alchemy-symbols-explained/

A Guide to Alchemy – The Science, Symbols, Elements, Books, and More. (2019, January 11). The Last Temptation of John. https://thelasttemptationofjohn.wordpress.com/2019/01/11/alchemy-symbols-elements/

Alchemy - Seven Stages of Alchemical Transformation. (n.d.). Www.deeptrancenow.com. https://www.deeptrancenow.com/exc3_7operations.htm

Alchemy Spagyric and Quintessence: spagyric remedies. (n.d.). Erboristeriacomo.it. https://erboristeriacomo.it/en/blog/139_alchemy-spagyric-quintessence.html

Cowie. (2018, February 27). Spagyric Secrets of The Alchemists: Alchemy as Alternative Medicine. Ancient-Origins.net; Ancient Origins. https://www.ancient-origins.net/history-ancient-traditions/spagyric-secrets-alchemists-alchemy-alternative-medicine-009655

Elemental Alchemy Symbols and Meaning on Whats-Your-Sign. (2018, January 31). Whats-Your-Sign.com. https://www.whats-your-sign.com/elemental-alchemy-symbols.html

Herbal Alchemy on a Budget - Alex Sumner. (n.d.). Www.jwmt.org. http://www.jwmt.org/v1n9/prima.html

How to Know When Your Elements Are In Balance Through Alchemy. (n.d.). Mystical Motherhood. Retrieved from https://mysticalmotherhood.com/how-to-know-when-your-elements-are-in-balance-through-alchemy/

Intrinsic.Roots. (n.d.). Spagyrics: An Introduction. AlcheMycology. http://www.alchemycology.com/spagyrics-an-introduction/

Koren, M. (n.d.). Alchemy Phase 1: Calcination. Spirit in Transition. https://spiritintransition.com/alchemy-phase-1-calcination/

McBride, K. (2018, January 29). Alchemy Oil: How to Make Herb Infused Olive Oil. Kami McBride. https://kamimcbride.com/herbal-alchemy-oil/

Plant Alchemy Archives. (n.d.). Apothecary's Garden. https://apothecarysgarden.com/category/astrodynamics/plant-alchemy/

Quintessence (alchemy). (n.d.). Www.symbols.com. https://www.symbols.com/symbol/quintessence-(alchemy)

The 22 Key Alchemy Symbols and Their Meanings – AxnHD. (n.d.). https://axnbc.com/2020/11/16/the-22-key-alchemy-symbols-and-their-meanings/

The Seven Basics Spagyric Tinctures. (n.d.). Quintessential Alchemy. https://www.quintessentialalchemy.com/the-seven-basics.html

Why Learn Real Alchemy on a Magical Path? (2016, September 27). Transcendence Works! https://www.transcendenceworks.com/blog/real-alchemy-magical-path/

Practical Guide to Alchemy and its Modern Day Uses. (2019, October 30) Reality Sandwich. https://realitysandwich.com/practical-guide-to-alchemy-and-its-modern-day-uses/

CPSIA information can be obtained
at www.ICGtesting.com
Printed in the USA
LVHW060910080521
686352LV00016B/585